THE EVOLUTIONARY BIOLOGY PAPERS OF ELIE METCHNIKOFF

BOSTON STUDIES IN THE PHILOSOPHY OF SCIENCE

VOLUME 212

THE EVOLUTIONARY BIOLOGY PAPERS OF ELIE METCHNIKOFF

Edited by

HELENA GOURKO

Belorussian University & Boston University, U.S.A.

DONALD I. WILLIAMSON

Liverpool University, United Kingdom

and

ALFRED I. TAUBER

Boston University, U.S.A.

SPRINGER-SCIENCE+BUSINESS MEDIA, B.V.

A C.I.P Catalogue record for this book is available from the Library of Congress.

ISBN 978-90-481-5356-5 ISBN 978-94-015-9381-6 (eBook)
DOI 10.1007/978-94-015-9381-6

Printed on acid-free paper

Printed and bound in Great Britain by MPG Books Limited, Bodmin, Cornwall.

Dedicated in admiration and friendship to
Robert S. Cohen and Annie Kuipers

TABLE OF CONTENTS

PREFACE

We are pleased to offer the English-speaking world the key evolutionary biology papers of Elie Metchnikoff (1845–1916), a work demanded by the rekindled interest in this key nineteenth century biologist. While hardly a household name today, Metchnikoff was a celebrated and well-known biologist, whose life has been well-documented. He was a founding father of immunology, and became a popular figure in France in the first decade of this century through his phagocytosis theory – an immune theory applied broadly to issues of health and disease. With the Nobel Prize for Physiology or Medicine co-awarded to him and Paul Ehrlich (1908), Metchnikoff's prominence was assured. The publication by his wife, Olga, of *The Life of Elie Metchnikoff* (1919) shortly after his death has long served as the official rendition of his career. The account, at one level, is a simple offering with a minimum of psychological and scientific complexity. Written as a personal retrospective, the logical development of the research that led to Metchnikoff's immunological theory is defined, and the thread of its subsequent defense is given with self-righteous zeal. In this sense, the biography presents an insightful glimpse as how Metchnikoff viewed himself: a figure whose logical approach, masterful defense, and honest intelligence attempted to devour, very much like the phagocyte he championed, his scientific opponents and establish the primacy of his research. As such, the evolutionary biology in which he was immersed during the two decades preceding his studies in pathology and host immunity were relatively minimized. But the significance of this earlier research can hardly be over-emphasized in terms of its importance in formulating Metchnikoff's better known studies in natural immunity.

With the recent publication of Daniel Todes's account of the historical context of Metchnikoff's research, *Darwin without Malthus, The Struggle for Existence in Russian Evolutionary Thought* (Oxford, 1989), and two intellectual biographies – the monograph, *Metchnikoff and the Origins of Immunology. From Metaphor to Theory* (Oxford 1991) by Alfred I. Tauber and Leon Chernyak, and the doctoral dissertation, "The wisdom of cells: The integrity of Elie Metchnikoff's ideas in biology and pathology" (Notre Dame University, 1991) by Craig Stillwell – students of nineteenth century Darwinism, developmental biology, immunology, pathology, and "socio-biological" thinking now have ready access to critical introductions to a corpus of research that clearly articulated the biology of that era and helped formulate its agenda. Here we offer the most important primary sources in their first English translation.

This translation project is the synergenic result between a translator/philosopher (Helena Gourko), a marine biologist (Don Williamson), and a historian/philosopher of science (Alfred Tauber). In our respective efforts, we have each profitted from the expertise of the other collaborators, and in framing this book, much has been appreciated heretofore ignored or unappreciated in Metchnikoff's opus. We can only hope that the reader also will profit from these diverse perspectives.

Our translation is based on Metchnikoff's Russian edition (*Academic Collection of Works* / Akademicheskoe sobranie sochinenii / vols. I-XVII. Moscow, Academy of

Medical Sciences of the USSR). As listed in the Table of Contents, some of these papers appear in the German and French literatures.

Two editorial points:

(1) Although Metchnikoff's style is complex – lucid, terse, and literary – the very density of the problems analyzed, as well as the significant difference between the expressions of sophisticated scientific and philosophical ideas in English and Russian, and more, the somewhat "archaic" nineteenth century Russian used by Metchnikoff make his writings difficult to translate. It is important to note that when Metchnikoff himself translated his works from Russian into other languages (he made a good many translations of this sort into German and French), he often transformed and changed them significantly. These changes did not alter the conceptual or descriptive details, but were obviously made for linguistic reasons. This immediately raises a major question concerning the degree of the translator's freedom, which, on the one hand, should be capable of keeping the content and context true to the original, but, on the other hand, must bring into consideration the significant inter-linguistic differences, as well as Metchnikoff's own approach in dealing with the problem. To resolve this difficulty fully is probably impossible, but it can be diminished by an understanding of Metchnikoff's major ideas and general orientation, which guided us in adjudicating any inconsistencies.

(2) Significant difficulties were created by extracting sections from the extended text of the "Essay on the Question about the Origin of Species" (238 pp. in printed form). Because of the limited space for this translation, this text has been reduced by a careful selection of the most relevant passages that emphasize Metchnikoff's relationship to the Darwinian debate.

Since Metchnikoff's views are not widely known, we have provided a concise introduction (based on Tauber and Chernyak, 1991), which presents the major ideas of Metchnikoff's views on evolutionary biology, as well as the results of his studies. The perspectives offered here concerning the development of Metchnikoff' career and his underlying philosophy are, to be sure, still open to scholarly debate and represent a summary of a previously presented interpretation.

Finally, without the encouragement of Robert S. Cohen and Annie Kuipers, this book would, in all likelihood, not have been completed. We thank them for their steadfast commitment to publish this translation. We, and the community of scholars at large, are in their debt. In appreciation of their friendship and support, we dedicate this volume.

Helena Gourko
Donald I. Williamson
Alfred I. Tauber

Boston University and The Isle of Man

June, 1999

INTRODUCTION

Why read Elie Metchnikoff's evolutionary biology papers? First, this research is of historical importance in its own right, reflecting a highly creative and critical opus of the latter half of nineteenth-century biology. Any consideration of this topic requires acquaintance with his thought. Second, although best known for his immunological theories, not his earlier work in embryology, it was Metchnikoff's foundational research in the problematics of Darwinian theory that drove his studies in pathology and natural host defense.

Metchnikoff (1845–1916) was a celebrated scientist, known best as a key founder of modern immunology, but he played a crucial role during the formative period of evolutionary biology by first helping to establish comparative embryology as important support for Darwinism, and then applying his notions of the organism to the new field of infectious diseases. His ideas concerning development were extended to adult animals in what he called the "phagocytosis theory," which became a critical component of a new discipline, immunology. As a system of defense in the struggle of competing species, Metchnikoff's "phagocytosis theory" extended the particular clinical concerns with infection to a grand biological theory. Thus his contributions to zoology, comparative embryology, evolutionary biology, comparative pathology, medicine, immunology, anthropology, and social theory should be regarded as of one piece, each aspect of his work contributing to the expression of a wider philosophical understanding of biology. Therefore, Metchnikoff can only be understood by integrating his thought, and thus his immune theories must be appreciated in the context of his broader contributions to evolutionary biology, which explicitly or implicitly served to frame many critical areas of twentieth century medicine (Tauber and Chernyak 1989) and biology (Tauber and Chernyak, 1991; Tauber, 1991).

Metchnikoff's primary scientific reports have remained largely untranslated into English. Of the many volumes of his published papers (seventeen in a definitive Russian edition), only a small fraction of these are available to the English-speaking reader. The best known translated works are his great summary texts, which are obviously valuable in serving as vehicles for his grand theories of biology and more general philosophical precepts, but these were primarily written as weapons in his polemics with other scientists. For instance, *Immunity in Infectious Diseases* (1901) is an elaborate argument for the centrality of Metchnikoff's phagocytosis theory against the competing views of German immunochemists. Such retrospective narratives of his work are inadequate records of Metchnikoff's research, the basis of his wider thinking, and the intimacy of revealing how a great nineteenth century zoologist regarded the major research issues of his time. In this regard, Metchnikoff's primary research is invaluable.

We have assembled and translated Metchnikoff's key evolutionary biology papers dating from his earliest writings (1865) to those of his mature period of the 1890s, endeavoring to produce a volume that will serve both as an introduction to his work for those interested in the broader reaches of his thought, and as a readily available source for scholars concerned with the particular historical development of evolutionary biology associated with his research. We believe that if these primary texts were more readily available, Metchnikoff's contributions to the development of evolutionary biology would better reveal the contested issues of the period and his own role in that Darwinian debate.

I

Metchnikoff's approach to Darwinism was highly convoluted and underwent several shifts. In his earliest writings he was critical of Darwin, but later in his career he became a sympathetic critic and finally an advocate. These postures were complex and oriented by the particular controversies in which he was embroiled. Suffice it to note here that his early attitude towards Darwinism was largely determined by what he thought to be its inability to elaborate proper terms for describing the difference between progress and regress, a description he deemed impossible in terms of adaptive evolution and natural selection. Ontogeny in his early view did not play any role in the Darwinian explanation of evolution through natural selection. In this respect, Lamarckian explanation of evolution, where an organism's activity played a much more important evolutionary role, was, from Metchnikoff's point of view, closer to the goals of morphology. (This position however did not exclude Metchnikoff's highly critical attitude toward Lamarck.) This general position seemed to dominate his thinking until after the phagocytosis theory was elaborated (ca. 1882), and then in 1892, when he published his essay "The Struggle for Existence Between the Parts of Animal Organisms," he explicitly proclaimed that the phagocytosis theory had been built upon the Darwinian theory of natural selection. But the true shift in his thought was catalyzed by his aggressive attack against Haeckel's *gastraea* theory ("The Struggle for Existence in a Broad Sense") in 1878. Metchnikoff's alternative to Haeckel's *gastraea* theory – the *parenchymella* theory – became the ultimate concretization of Metchnikoff's general evolutionary studies and the primary ground from which his phagocytosis theory arose.

Metchnikoff's immunological theories were conceived as a response to answering the challenge of what served to regulate and define organismic growth in a biological universe devoid of essentialism. His response, immunity, was an answer coming at the end of a long line of research in the apparently different domain of embryology. Thus Metchnikoff's immunological theory arose from a deep commitment to problematics of evolutionary biology and in a sense represent the maturation of his thought despite the apparent change in venue. The general problem of the struggle for existence and natural selection was transformed into the concrete and specific question of the nature of inflammation. These evolutionary biology papers serve as the conceptual foundation for Metchnikoff's immunological texts, which are better known than these earlier responses to the Darwinian question. So, we might best understand the continuity of Metchnikoff's thought as tied to the fundamental challenge of the organism's identity in the context of Darwinism.

Darwinism had surely the most important influence on Metchnikoff's scientific and metaphysical development. Metchnikoff read *On the Origin of Species* (1859) in 1863 (then eighteen years old), and the review he wrote at that time ("A Few Words About the Modern Theory of the Origin of Species") was largely unfavorable. His relationship to Darwinism remained critical for approximately 20 years, but after the mid-1880s, Metchnikoff's intellectual posture grew more closely akin to the natural selection theory, and his retrospective musings of his earlier negative attitude towards Darwinism was glossed over. During the post-1880s reaction, a period of Darwinism's eclipse, Metchnikoff professed himself to be an ardent supporter.

It is important to note that Metchnikoff came to agree with Darwinism not as an ideology (or at least not as an ideology only), but as a scientific theory. Metchnikoff's two early essays, "On the Developmental Life-History of *Myzostomum*" and "On the Embryonic Development of Lower Crustacea" (both published in 1866 and included here), are very important in this respect. Not referring directly to Darwinism, these papers investigate problems of morphology of lower animals according to a concept of transformation (although still quite critical). There is no doubt that starting with these texts (and with Metchnikoff's abbreviated summary of Haeckel's *Generelle Morphologie* published in 1869), his arguments against Darwinism were voiced as a morphologist, with a gradual change of accents. His embryological studies of the period after 1872 witnessed a radical change in his view of the role in which the genealogical perspective might play in comparative embryological studies. Previously he regarded the task of phylogenetic reconstruction as no more than an application of comparative embryology, which could not be particularly helpful in the pursuit of its own proper goals. But by 1876, Metchnikoff claimed that his view of morphology was, in principle, evolutionary and that morphology was determined by genealogy. A theoretical elucidation of this opinion, as well as its experimental support are included in a series of Metchnikoff's works grouped under the title "Comparative Embryological Studies" (1881–1885), and "Embryological Studies on Medusae" (1886).

In 1876, Metchnikoff published his definitive work on the problem of transformism: "Essay on Questions about the Origin of Species". This essay directly refers to the importance of genealogy for morphological research. Here Metchnikoff claimed that his view of morphology was, in principle, evolutionary and that morphology was determined by genealogy. The last section of this essay explicitly demonstrates, however, that he undertook the investigation of evolutionary theory for the sake of his own central research pursuit, namely the identification of the primary embryological layers, in general, and of the mesoderm, specifically. The design of the entire paper was to focus on Metchnikoff's argument against Haeckel's *gastraea* theory, but that critique does not prevent Metchnikoff from a much broader consideration as to what degree (if at all) it was possible to describe developmental processes in Darwinian terms.

This work was a continuation of his previous attempt to generalize Darwinian ideas within the context of anthropology ("Anthropology and Darwinism", 1875). It was clearly a pioneering approach to the application of Darwinism to neighboring disciplines, an exploration designed to test the applicability of major Darwinian ideas beyond their narrow purview. (Metchnikoff's works on anthropology and ethnography comprise a separate volume of the Russian edition, #16). This essay is important also for understanding Metchnikoff's philosophical evolution which later resulted in his "philosophy of optimism".

The last of the Metchnikoff's texts included here is the essay, "The Struggle for Existence Between Parts of the Animal Organism" (1892), in which he expounds his favorite theme, the inherited disharmonies of organisms and the central physiological role of phagocytes. Besides its importance for presenting Metchnikoff's final acceptance of Darwinism, it is interesting as a continuation of the bitter debate between Metchnikoff and Haeckel. The title of this essay is almost identical to a book published in 1881 by Wilhelm Roux, a student of Haeckel. But the shared titles hardly reflect an essential agreement of opinions concerning organismic integrity. Metchnikoff's view of disharmonies and Roux's teaching of the struggle between the parts of the organism had a different philosophical foundation and a different scientific relationship to Darwinism. Roux's book is an expression of orthodox Darwinism, while Metchnikoff's essay presents a significant modification and re-thinking of Darwinism in conjunction with his own phagocytosis theory. The extended argument was published by Metchnikoff as *Lectures on the Comparative Pathology of Inflammation.*

Following a brief biographical orientation in the next section, the intellectual context of these selections will be considered.

II

Metchnikoff was born on May 16, 1845 in Kharkov Province, Russia, the youngest of five children in a privileged Russian family. He was tutored at home, and his volatile nature earned him the nickname, Quicksilver, which might well have been retained in his later years during the highly defensive and impassioned defense of his scientific views – first as a comparative embryologist, later as an immunologist. He was energetic, flamboyant, and manipulative. In 1856, he enrolled in the Kharkov Lycée. An excellent student, he pursued his early interest in science aggressively, attending supplemental lectures at the university and beginning his microscopic studies of infusoria. Already in his adolescence, being attracted to Rudolf Virchow's cellular theory, Metchnikoff had visions of creating a grand theory of medicine. His first acknowledged publication was a critical review of a geology textbook published in the *Journal de Moscow.* Metchnikoff was sixteen years old. The facts point to singular purpose, precocious intelligence, and highly energized motivation. But sensitivity and naiveté took its toll for the *wunderkind.* Attracted to Germany, he wished to enroll at the University of Wuerzburg and study with Rudolf Koelliker, but after suffering adjustments problems, Metchnikoff quickly returned to Kharkov and instead enrolled at the university there. The most important result of his visit to Germany was the acquisition of Charles Darwin's *Origin of Species,* a seminal work for Metchnikoff's personal scientific career.

Metchnikoff embarked on his formal zoological studies under the tutelage of Ivan Petrovich Tshelkoff, one of the few Kharkov faculty he respected. Not content with attending lectures, Metchnikoff began his own research with a study of the mobile stalk of a ciliated infusorian, the *Vorticella*, to characterize whether it was analogous to muscle in higher organisms. His paper, published in *Muellers Archives* (1863), was severely criticized by Wilhelm Kuhne, and after Metchnikoff confirmed his original conclusions (1864), Kuhne rebutted. The bitterness of this first encounter repeatedly punctuated Metchnikoff's career.

Metchnikoff's formal university education lasted only two years, condensing the usual period by half. He aborted his research activities, intensified his studies with minimal social distractions, and thus passed his qualifying exam. He still required a thesis, so in 1864, the as yet unfocused student went to the North Sea island of Heligoland, where the botanist Ferdinand Cohn guided him to Rudolf Leuckart of Giessen University. Metchnikoff read two papers at the Naturalists' Congress held in Giessen; struck by his precociousness, Leuckart immediately accepted him into his laboratory. The two key events of this period were reading Fritz Mueller's *Fuer Darwin* (1864) and discovering the alternation of sexual and asexual generations in nematodes, wherein parasitic hermaphroditic worms give birth to a free bisexual generation. According to Olga Metchnikoff, Mueller's book had a profound influence because it was the first to purportedly confirm Darwin's theory by tracing genealogies through embryonic structures. However, we view the true basis of Metchnikoff's research prior to the mid-1870s with some skepticism in regard to this grand scheme of confirming Darwinism. This issue will be discussed in detail later, but it suffices to simply note here that because of eyestrain, Metchnikoff took a vacation (visiting his brother in Geneva) and discovered that Leuckart had stolen his results and published them independently. Metchnikoff publicly denounced the theft (in DuBois-Reymond's journal [1865]) and in the same year went off to Naples with a grant from the Russian Ministry of Public Instruction.

Metchnikoff's singleness of purpose must have insulated him from both professional frustrations and ill health. In Italy, he began a systematic study of the germ layers in invertebrate embryos, sharing common interests with another Russian zoologist, Alexander O. Kovalevsky, with whom he had a long-standing and productive (at times acrimonious) collaboration. In the autumn of 1865, Metchnikoff continued his formal embryological studies in Goettingen and then Munich, independent of his faculty mentors. Once again, he went to Naples in 1867, where he and Kovalevsky finished their studies. They returned to St. Petersburg to receive their doctorates and share in the prestigious von Baer Prize.

At the age of twenty-two, Metchnikoff was appointed *docent* at the new University of Odessa where he remained until 1868, when he moved to St. Petersburg. His scientific research of this period was conducted on sponges and echinoderms, but he was bitterly disappointed with his working conditions, salary, and general disposition of life in St. Petersburg, which reinforced his growing melancholy and pessimism. Metchnikoff returned to Odessa in 1872 and remained a member of the university faculty until 1882. Metchnikoff's Odessa period (1872–1882) was marked by an active teaching schedule and involvement in academic politics. There was feuding between the Ukrainian and Moscow faculties along liberal-conservative ideologies. Although attempting to remain aloof, political events caught him in their maelstrom. With the assassination of Tsar Alexander II in 1881, reactionaries gained control, limited the independence of the university, and made political appointments to the faculty. Matters worsened and Metchnikoff was active in the attempt to restore academic integrity; in the process, he witnessed the collapse of academic freedom and resigned.

So, by 1882, while in Messina, having escaped from the political turmoil of Odessa, Metchnikoff concentrated his scientific efforts on establishing the basic function of mesodermal amoeboid cells through the phyla, thus merging his interest in embryology with a newly acquired commitment to inflammation. This is where the popular biography of Metchnikoff usually begins. A scientific fantasy, first told by Metchnikoff himself and uncritically repeated, has been offered to account for a sudden insight, which oriented Metchnikoff's theory of immunity for the next thirty years. The story goes that Metchnikoff was conducting studies in the parlor of their Messina apartment, and to gain some privacy he sent the family off to the circus. While away, Metchnikoff pierced star fish larvae with rose thorns obtained from the garden, and the next day saw phagocytic cells surrounding the intruders. Metchnikoff deduced that the cells were attempting to devour the invaders and through their intrinsic digestive capacity, a capability preserved throughout phylogeny, animals with a gut had at hand a new functional element. No longer required to digest devoured particles for nutritive purposes, these phagocytic cells were poised and ready to "eat" micro-organisms and thus serve as a defensive arm against pathogens. Thus "phago" (to eat) – "cyte" (cell) was coined.

However this episode was not the flash of genius attributed by a romantic account, but an important stage along a complex road reaching back to Metchnikoff's earliest embryological studies, where he was attracted to amoboid cells as "markers" of the mesoderm. In a highly creative fashion, he discovered that these cells possessed multitudinous functions, which was elaborated into his theory of phagotosis. As early as 1865 in Giessen, Metchnikoff studied the terrestrial flatworm (planarian) *Geodesmus bilimeatus*, which lacked a gut and whose digestion he compared to the digestive process of protozoans (1866). Between that study and his extensive research begun around 1877 on intercellular digestion of invertebrates, he was primarily preoccupied with the search for fundamental principles to establish a theoretical basis for comparative embryology. This line of inquiry was then translated into the predominant interest in inflammation and infectious diseases. [1]

Metchnikoff 's early work, in a sense, predicted his scientific development, and we may well appreciate the inner dynamics of his thinking from this period, which we will detail below. In short, Metchnikoff underwent a metamorphoisis, of sorts. His later evolutionary approach to embryology (after the mid-1870s) excluded the assumption that in ontogenetic development there was an enactment of some universal, predetermined scheme of evolution. This was in direct conflict with Ernst Haeckel and other prominent embryologists of the time. In contrast, Metchnikoff projected embryological data to offer a powerful means for a reconstruction of some particular, as opposed to general, phylogenetic development. This issue focused his last studies in embryology, which were eclipsed by his almost total commitment to research in immunology by the mid-1880s.

With the death of his wife's parents in 1881, Metchnikoff was endowed with financial resources that allowed him to move to Italy to conduct his research free of academic and political encumbrances. He initially carried out his pathology experiments in the realm of physiological inflammation and continued his embryological research with the formal presentation of the *phagocytella* primal metazoan (earlier called *parenchymella*) in 1886. But Metchnikoff soon became a "pathologist", abandoning his embryological orientation altogether by the early 1890s. It is of particular interest that this career shift coincided with the emergence of experimental embryology in the same period, an avenue he never pursued. Strictly from a methodological perspective, it was quite simple for Metchnikoff to apply his extraordinary observational skills to the newly developing field of immunology, where he remained an "observational" biologist, leaving the emerging field of immunochemistry to others.

By 1886, Metchnikoff had become an international scientific celebrity. He was invited back to Odessa to head a research institute (similar in design to the Pasteur Institute) devoted to bacteriological studies. This was his last Russian appointment. His desire to emigrate was sealed by his frustration owing to internal political and administrative intrigue, by his inability to conduct a human immunization program (because he was not a physician), and by a disastrous innoculation program of sheep against anthrax. Metchnikoff visited Germany, looking for refuge, but was either received with hostility or otherwise found his options limited. Fortuitously, he visited Louis Pasteur, who greeted him warmly. The Pasteur Institute opened in 1888, and Metchnikoff was given an entire floor for his cellular immunology research. He remained at the Institute until his death on July 16, 1916. The first half of this last period of his career was devoted to a ferocious defense of his phagocytosis theory and is conveniently bracketed by two summary texts: *Lectures on the Comparative Pathology of Inflammation*, derived from lectures he delivered at the Pasteur Institute in April and May 1891 (1893, 1968), and his consummate opus, *Immunity in Infective Diseases*, published in 1901 (1905, 1968). These works present fair and comprehensive statements of Metchnikoff's scientific posture of this period: The phagocyte was the defensive moiety of immunity, with humoral factors but elaborated phagocyte products. In 1908, Metchnikoff went to Sweden to accept the Nobel Prize and then made a triumphant tour of Russia. His role as elder scientific statesman limited his scientific production and the last years of his life were marked by poor health.

III

When Metchnikoff began his research in embryology, the embryonic layer concept had almost a fifty-year history. Christian Pander first described the trilaminar structure of the earliest incubated chick embryo stage (1817), which was developed further by Ernst von Baer, Pander's friend and colleague, who was the true founder of comparative embryology [2]. Von Baer expanded Pander's germ-layer concept for the chicken to all vertebrates (1828–1837). From his own studies, von Baer concluded that every layer contained within itself a germ of a set of organs, which developed in accordance with a single general plan *(Bauplan)*. According to von Baer, the first layer is the origin of the "animal life" organs (the organs of motions and the nervous system); the second layer, the organs of "vegetative life" (the digestive organs); and the third layer, the blood vessels. Von Baer observed formation of the digestive and nervous tubes and the somites as the basic elements of primary development. He described the provisional embryonic membranes and began study of their role in embryonic development.

A new stage in the history of embryology followed Theodor Schwann's establishment of the cellular theory in the late 1830s, which asserted cells as the basic element of living tissue, and by the 1840s, the histological approach to the embryological layer problem was elaborated. Rudolf Albert von Koelliker, in his 1842 dissertation "De prima insectorum genesi," concluded that there were analogies between the development of arthropods and of vertebrates. In 1849, Thomas Huxley noted an essential similarity between the two primary embryonic layers of vertebrates and the two layers structuring the adult forms of Coelenterata (1849) [3]. Generalizations by Koelliker, Huxley, and Remak, although they appear from our perspective to be purely empirical and descriptive, were supported by specific natural philosophical speculations concerning the universal unity of the animal realm, an intellectual relationship not lost on Metchnikoff (1866) [4].

Once the ovum was recognized as a single cell, multiple divisions that formed the primary layers were shown to follow fertilization. Prior views of the embryonic layers were reconsidered from this new histological vantage, first by Robert Remak in 1850–1855, who recognized three layers, but, at the same time rejected von Baer's interpretation. He discerned the layers according to their future histological development instead of serving as germ carriers of future "animal-life" and "vegetative-life" organs as previously regarded. Remak recognized *two* primary embryonic layers: the upper one subdivided into medullar and epidermic plates; the lower layer, in turn, divided into two sublayers: the trophic layer, which forms the alimentary canal and its derivatives; and the middle layer, which forms muscle, connective tissue, blood vessels, sex glands, and peripheral nerves. The middle layer was thought to further divide into dorsal and ventral somite plates by a cavity, which we know now as the coelom.

After Remak's vertebrate embryology studies, which established the histological approach, Zaddach undertook research in insect embryology, where the attempt to draw an analogy with vertebrates also played an essential role (1854). He described a split of the so-called embryonic stripe into two layers that corresponded, not (as accepted earlier) either to the serous layer (the middle layer of Remak's schema), or to the mucous layer (Remak's inner layer), but to the horn (external) and muscle layers of vertebrates. Besides this parallel, Zaddach believed that he had also found other similarities, being homologous with the primary segments of arthropods and the plates of the vertebrates. Zaddach's opinions were broadly supported – for example, by Huxley (1858) and by Rudolf Leuckart (1858). Metchnikoff cited Remak as the most prominent figure in animal embryology of the 1850s (Metchnikoff, 1869). While most of the Remak's findings and interpretations agree with our current concept of embryonic layers (except for the origin of the peripheral nerves), most interesting in understanding his works is a parallel between Remak's concept of the layers and Metchnikoff's ideas, developed, not in the middle 1860s, but in the 1870s and 1880s.

If in the middle 1860s there was obvious progress in the elucidation of vertebrate comparative embryology, the state of invertebrate comparative embryology was quite different. Darwin's *Origin of Species* dramatically stimulated a change in this situation, offering a new meaning to the program of establishing "a general affinity between separate animals." Note Metchnikoff's appraisal written in 1869:

Most studies on external forms of development paid comparatively very little attention to the formation and development of organs. Because of all of this, purely comparative embryological questions have been laid aside, not only without elaboration but without even being formulated. With this state of affairs, all the accumulated material could not be put together in a single whole but was destined to exist in the shape of an unsorted mass. And meanwhile the goal of the history of development, a science which deals with the multitude of changing forms, must be to establish precisely the general affinity between different animals and to search for the plan of their organization. (Metchnikoff, 1869)

The key early texts addressing this problem were Fritz Mueller's *Fuer Darwin,* published in 1864, and Ernst Haeckel's *General Morphology of the Organisms* (1866). Metchnikoff's intensive work in this field started in 1865; his friend Kovalevsky, who was five years older, began his research in 1862, and together they must be regarded as the cofounders of invertebrate comparative embryology. The received view, derived from Metchnikoff himself, as well as other commentators, was that from an early beginning, Metchnikoff was devoted to Darwin's theory, which focused Metchnikoff's own interests in zoology. As already noted, sometime around 1865, he read Mueller's *Fuer Darwin*, and, according to Olga (Metchnikoff 1919, p. 50), this book had a decisive influence on his career. Working with crustaceans, Mueller had demonstrated that the most important insight concerning animal genealogy was to be found in their embryology, a basic hypothesis that supported the future formulation of the so-called Mueller-Haeckel biogenetic law.

The 1860s were the crucible of Metchnikoff's career, and it was then that the various paths he was to follow were first introduced. While extensive knowledge of the three primary embryonic layers in vertebrates had been accumulated, correspondingly little was known about invertebrates, and it was here that Metchnikoff began his first zoological studies with his friend Kovalevsky. In 1865, Kovalevsky discovered the development of two embryonic layers in *Amphioxus* (lancelet) larvae. Having observed division of the lancelet's fertilized ovule into a multicellular sphere -the blastula – he then saw that one-half of the blastula sinks into the other half (invagination), thus forming a two-layered creature – the gastrula. Soon thereafter, the surface of the embryo is covered with cilia, and the oval embryo ruptures the ovule's membrane and frees the gastrula to begin its free-swimming phase. Lancelet, a chordate, has larvae that are similar to those of invertebrates, and Kovalevsky extended these observations to other simple metazoans, obtaining similar results. Extrapolating from these findings, he believed that invagination, the forming of two embryonic layers, was characteristic for most, if not all, multicellular organisms.

Metchnikoff was obviously impressed with Kovalevsky's findings, but quickly extended and challenged them. In 1865, Metchnikoff discovered among sponges, hydroids, and lower medusae a second pattern of embryonic layer formation, observing that the second layer (the endoderm) was formed not by means of invagination (of the ectoderm), but by introgression – migration of a number of flagellated cells from one pole of the blastula wall into the central cavity. These cells drew in their flagella and became amoeboid and mobile, multiplied by division, filling the cavity of the blastula. These cells were capable of digestion, and were correctly designated as primordial amoeboid, phagocytic cells, the central actors of his later immunological studies (Metchnikoff 1893, 1968). Metchnikoff named this stage, *parenchymella*, a primitive organism that would later contest Haeckel's own hypothetical first metazoan model creature. Metchnikoff observed in those organisms forming a digestive cavity that this parenchymatic mass differentiated further into two layers: the mesoderm and the endoderm. The key finding, however, the one that would figure prominently in later polemics concerning the evolution of metazoans, resided squarely on this different mechanism of gastrula formation, namely, introgression, as opposed to invagination.

Haeckel, using Kovalevsky's discovery of the development of a lancelet larvae, formulated his famous *gastraea* hypothesis based on the invagination mechanism and extended that finding to a sweeping hypothesis that had the common metazoan ancestor as developmentally similar to the lancelet's gastrula (Haeckel, 1872). This model was supposed to provide comparative embryology with a phylogenetic basis. Acknowledging the importance of this kind of theoretical reconstruction, Metchnikoff, on the one hand, argued that the true author of the hypothesis was Kovalevsky. On the other hand, he asserted that his own *parenchymella* more closely fitted the role of the embryonic image of the primordial metazoan, inasmuch as the more primitive mechanism, as seen in lower organisms, was introgression. Haeckel's model presupposed that extracellular digestion within a digestive cavity was the basic feature of complex organisms, recapitulating the function of the *gastraea* digestive cavity. Metchnikoff's *parenchymella* represented intracellular digestion as the common feature of unicellular organisms and most primitive metazoans.

This archaic form of digestion had been preserved in the ability of the amoeboid mesodermal cells of the higher animals to perform intracellular digestion – the basic mechanism Metchnikoff later argued was the basis of vertebrate host defense. Thus, according to Metchnikoff, the preserved function of these mesodermal cells provided an evolutionary basis for the comparative embryology of animals. So, beyond the phylogenetic reconstructions study of these cells would offer, Metchnikoff was to build an entire theory of immunity based upon the diverse function of such amoeboid, phagocytic cells. The study of the evolutionary fate of this function led Metchnikoff to the dual ideas of phagocytosis as the crucial process in ontogenetic development and as the primary mechanism for active immunological defense. This is the prehistory of the phagocytosis idea as it was seen by Metchnikoff himself and repeated by many historians. But this recount is a retrospective reconstruction from the perspective of Metchnikoff, the committed Darwinian and leading immunologist, as opposed to Metchnikoff, the critical evolutionist, working before his Darwinism matured and his defense of his immunological thesis was fully articulated. Thus there are glaring inconsistencies in Metchnikoff's later recounts as attested by the documents from his early career (scientific publications, manuscripts, and private correspondence), which fall among the general issues concerning his relations to Darwinism, and the positions held by Mueller, Kovalevsky, and Haeckel:

(a) Metchnikoff read Darwin's *Origin of* Species in 1863, but, as seen in his essay written in the same year (1863b), his reaction to the book was highly critical. He did not change his negative position for the several years.
(b) Metchnikoff read Fritz Mueller's *Fuer Darwin* around 1865, but this text hardly determined his approach to comparative embryology. Even in 1869, four years after Kovalevsky's description of the gastrula stage, when Metchnikoff himself was already deeply involved in comparative embryology, he firmly opposed those embryologists he called the Darwinians, and he included Mueller and Haeckel as prominent among them.
(c) In 1902, Metchnikoff wrote that in Naples during 1865 he was able to convince himself of the truth and importance of Kovalevsky's discovery concerning lancelet larvae development. But, in fact, during Kovalevsky's defense in 1865 of his master's thesis about the development of *Amphioxus* (lancelet), it was Metchnikoff who heatedly contested Kovalevsky's scientific claims. Metchnikoff at this time argued that the thesis was nonsense, for the digestive cavity could not be formed by invagination. In 1866, Metchnikoff published a paper in which he stated that Kovalevsky's assertion about invagination was in contradiction with all known facts (Metchnikoff 1866b). The polemic continued at least until 1873, and thus, it is problematic to refer to parallel and coordinated work of the two friends.
(d) In his publications immediately following Kovalevsky's discovery, Metchnikoff did not write about the importance of reconstructing evolutionary relationships, and indeed, he asserted quite the opposite: morphological similarities of different groups of animals are well-known, but there was no necessity to interpret them as a reflection of "blood relations" (1869). There is no evidence that in his early comparative embryological studies Metchnikoff considered this kind of reconstruction of any significant importance for comparative embryology itself. [5]

Metchnikoff did observe an alternative way of formulating primary embryonic layers in sponges, hydroids, and lower medusae, but those observations do not necessarily indicate that from the very beginning he considered those results as a basis for an alternative to Haeckel's phylogenetic reconstruction. Metchnikoff, in his later polemic against the gastraea theory, accused Haeckel of nonscientific naturphilosophical generalizations, and we might concur that in many respects Metchnikoff's hypothesis was more steadfast to the known facts of the period. But before the specific arguments ensued, Haeckel directed Metchnikoff to recognize the importance of phylogenetic reconstruction for studies in comparative embryology. The gastraea theory concentrated his attention on the role of intracellular digestion in phylogenetic and ontogenetic development. In short, the facts Metchnikoff proposed as the basis of his pa*renchymella hypothesis* had been discovered by Metchnikoff himself, but the logic that structured those observations into a hypothesis had been elaborated in the arguments with Haeckel. In short, Metchnikoff's hypothesis was a response to the problems generated by the gastraea theory, which had failed, at least to Metchnikoff's way of thinking, to offer a satisfactory solution to the question of metazoan primordial development.

These discrepancies between Metchnikoff's retrospective history of his research which resulted in the phagocytosis hypothesis can hardly be explained by his ambitious unwillingness to recognize his own mistakes or by his intention to affirm his own priority in every step of the process. The inconsistencies rather are the result of an attempt to logically develop the phagocytosis hypothesis, and from this tidied up retrospection, Metchnikoff presented a refined and altered attitude toward Darwin, Mueller, Kovalevsky, and Haeckel – a retrospection biased by the end point of his research career. How we might understand that work from another perspective follows.

IV

The debates concerning the evolutionary significance of the embryonic layers started in the second half of the 1860s with Kovalevsky's research (in 1865), who (as already mentioned) argued that the most general pattern of layer formation proceeded by several cleavages of the fertilized ovum, leading to formation of a primary multicellular sphere, followed by the production of a second layer by invagination. These two steps would then form a two-laminar sac. The third layer was thought then to arise from the cells of the first two layers. From Kovalevsky's (and after him Haeckel's) point of view, this manner of embryonic development reflected the most ancient stage of evolutionary formation of multicellular organisms. Metchnikoff admitted the importance of this kind of genealogical reconstruction only in the 1870s, when he initially participated in discussions of this matter, but he had already proposed another model for the formation of the primary layers. As already discussed, in his original model, the middle layer did not develop *after* formation of the outer and inner layers, but rather, the primary sphere (the blastula) gradually filled in with a parenchymatic mass of cells (by introgression) that further divided into the inner and middle layers (the endoderm and the mesoderm). Thus, Metchnikoff's concept of the embryonic layers (developed in the 1870s and 1880s) is homological relative to that proposed by Remak who (in the 1850s) also asserted that the inner and middle layers were formed by division of the second layer.

The dominant theme in Metchnikoff's reconstructed history is clear: The general interest concerned the problem of how to apply (if at all) the concept of embryonic layers to invertebrates. But now let us return to the genesis of his research in this area. The true origin of his interest in insect embryology – lost in the theoretical polemics and reconstructions of his later positions revolving around Darwinism – was inspired by a more particular challenge: Nikolay Wagner's discovery in the early 1860s of the phenomenon that was later defined by von Baer as pedogenesis (parthenogenetic reproduction by insect larvae structurally incapable of copulating). Wagner observed that certain Cecidomyiidae larvae spawned new offspring before attaining adulthood; these juvenile organisms developed within the mother larva, ultimately destroying it. The larvae are generated in this way in the fall, winter, and spring, but during the summer, the last larval generation matures into pupae from which appear the reproductive forms, the small flies. The females lay fertilized ova and the cycle begins again.

Metchnikoff's first embryological paper (prior to any research referable to the problem of embryonic layers) was devoted to this problem of Cecidomyiidae larva development (1865). Wagner argued that the larvae developed from the fat body cells; others disagreed, thinking that either the larvae developed from germ cells or from special germ organs. The latter were described by Leuckart as well as Metchnikoff, who worked in Leuckart's laboratory at this time. As the ovum was thought not to possess a distinct delimiting membrane, Leuckart called the ova (as opposed to fertilized, shell-covered ova) the pseudo-ova. Metchnikoff, in describing the development of the larva from the pseudo-ovum, made two important observations: (a) the nuclei of the blastoderm (the single-layered structure formed by the cleavage of the ovum) are formed out of the germ vacuoli (i.e., the ovum's nucleus), and (b) Weismann's so-called polar cells gave birth to the sex cells of the next generation. Ganin at Kharkov University also described these germ organs. In contrast to Leukart and Metchnikoff, he argued that the cells from which the larvae developed were not pseudo-ova, but true ova; he thus considered Wagner's process as nothing more than parthenogenesis. Thus arose a controversy that directed Metchnikoff to continue his studies of Cecidomyiidae. Writing to von Baer, Metchnikoff requested an arbitration of the dispute. Metchnikoff's letter was published (1866b) and von Baer in his notes to the letter and in a special paper devoted to the problem (1866) asserted that the true difference between typical parthenogenesis and the reproduction of Cecidomyiidae was not the question of true ova versus pseudo-ova, but rather that in true parthenogenesis the adult female laid ova, whereas Cecidomyiidae ova were laid by its larvae. Von Baer proposed the now-accepted term, pedogenesis. Metchnikoff considered his Cecidomyiidae paper and the one that followed (1866c) as preliminary reports, which he intended to include in a later work on insect embryology: "Embryological Research on Insects" was published in 1866. For this contribution, Metchnikoff shared with Kovalevsky the prestigious von Baer Prize in 1867.

Each of Metchnikoff's predecessors, Zaddach and Weismann, studied only representatives of one order of insects, but Metchnikoff's work encompassed representatives of Diptera, Hemiptera, and Homoptera. Fixed microsections had not yet been developed, and Metchnikoff could only observe living larvae. Therefore, he was limited in his studies to the development of external structures (but only in their early development) and to the emergence of internal organs (but only insofar as transparency allowed). This technical limit was the principal restriction in the studies of the role of embryonic layers in organogenesis. In fact, the very existence of invertebrate embryonic layers was unresolved for the same reason. Despite the methodological problems, Metchnikoff came to a number of important conclusions. We noted, for example, his observation that pole cells (discovered by Weismann) give rise to sex cells. In addition, the issue of blastoderm development into more complex structures had not yet been defined, and Weismann, who had proposed two lines of maturation – one involving a rupture of the blastoderm and the other leaving the blastoderm intact – was correctly refuted by Metchnikoff.

Working in 1865 in Leuckart's laboratory (devoted mainly to parasitology), Metchnikoff in parallel to his studies of insect embryology initiated a second theme of research: the development of *Ascaris nigrovenosa* (a nematode), a parasite of frogs, that has alternating parasitic and free generations. (Worthy of note, this was the first project involving parasites, an interest Metchnikoff returned to throughout his life.) In 1865, he studied the development and morphology of a crustacean, *Nebalia,* concluding that it was closer to the highest crustaceans, Malacostraca, then to the Phyllopoda as previously thought. (The studies of *Nebalia* composed his doctoral dissertation [1868]). Also in 1865, Metchnikoff discovered the larvae of *Balanoglossus* and in 1869 returned to this topic and proved that the earliest known tornaria larva was one of the earlier stages of the larvae of *Balanoglossus.*

This line of research continued work on the larvae of Chaetopoda (1869; published with the naturalist, Edouard Claparede). Their goal was to determine whether the classification of the diverse larvae proposed by Johannes Peter Mueller corresponded to the classification of adult forms; Metchnikoff continued related research until 1871. In 1869, Metchnikoff published his extant work on the development of Echinodermata and Nemertinea (his first publication on the topic appeared in 1868), an opus some specialists have considered especially important in the field. Metchnikoff correctly described each of the basic problems of echinoderm development (enterocoelom, metamerism, metamorphosis), and was the first to observe formation of the coelom in animals without metamerism in adult states.

So in summary, Metchnikoff's research on insect development was initiated by the question of pedogenesis, not the genealogical significance of embryological layers. Although he derived important results on insect embryology (including his observation of nuclear reproduction in the egg and superficial cleavage), the primary layers were only part of his concern. For instance, in this work (and in some others), he paid no less attention to the problem of embryonic membranes, apparently viewing these structures as highly important for establishing a similarity in the development of vertebrates and invertebrates. In his doctoral dissertation on the crustacean *Nebalia* (1868), Metchnikoff wrote little of the embryonic layers. As mentioned, the technology of the time limited discernment of layers in eggs that contained an extensive amount of yolk. Seeing two layers, he believed that the inner one was the serous layer (i.e., the endoderm), but beginning in 1868, with the studies of scorpion development, Metchnikoff applied the new techniques of fixation and cross sections, a technical innovation which was to revolutionize the discipline.

V

So now let us provide an overview of Metchnikoff's research during the first half of his career. As already noted, the most influential model in embryology during the 1860s was that proposed in the 1850s by Robert Remak, who, in accordance with the recently formulated cellular theory, gave a histological orientation to the concept of the primary layers. The model had been elaborated for vertebrates, and although by 1865 extensive material in the embryology of invertebrates had been collected, little was known about their layers, and Metchnikoff concluded that the comparative embryology of invertebrates had not yet reached maturity. But in the mid-1860s, August Weismann published his works on Diptera development in which, in tracing Zaddach's "horn layer," he came to the conclusion that it was not homologous to the horn layer of vertebrates (1863; 1864a; 1864b). Weismann viewed the "horn layer" that Zaddach had allegedly observed in insects as a special formation, which he named the "fold layer," concluding that nothing in insects corresponded to the embryonic layers of vertebrates, and as a result he rejected any parallel between the embryology of arthropods and that of vertebrates. Although Metchnikoff expressed skepticism about Huxley's and Leuckart's enthusiastic support of Zaddach's ideas, we may not infer that he totally rejected the scheme. After all, at the very time when Weismann was publishing his work on Diptera, Metchnikoff was working in Leuckart's laboratory, and he believed that, in rejecting any similarity of vertebrate and insect development, "Weismann fell into another extreme and in this respect he hardly took a more correct way. . . . I felt myself now forced to undertake further research in the so peculiar sphere of the embryology of insects, and to gain footholds which would help somehow to explain all these peculiarities." [6]

In the late 1870s, and especially after his proposal of the plagocytosis hypothesis in 1883, Metchnikoff formulated a version of his scientific opinions and achievements between 1865 and the mid-1870s. According to this revised version of his scientific development (until 1872), he was inspired by the idea of establishing evolutionary relations between different groups of animals. Pursuing this goal, and firmly believing in basic evolutionary unity, Metchnikoff's revisionism saw him defining, in different groups of animals, the same pattern of basic structure: the three embryonic layers. This story is broadly accepted in the Metchnikovian literature. [7] But what is the concrete record? During that period, Metchnikoff understood the concept as Remak had elaborated it for vertebrates. In his own studies of invertebrates, he sought similar structures. [8] In some instances, he was able to establish a trilaminarity (1871), in others he found only bilaminar structures (1867), and in both cases, he was not always certain of the correspondence between these structures and Remak's layers. [9] Occasionally, he failed to establish the presence of these types of structures at all. [10] In these circumstances, it was impossible to assert the universal presence of the layers and the validity of Remak's concept as applied to invertebrates. In short, from these data it was possible to conclude that although the formation of the layers is widespread among invertebrates, the similarity is not so great as to serve as the foundation for any morphological conclusions. The formation of the layers being different in different animals, denies deriving definite relations between groups of tissues. Thus, for example, the middle layer of Cephalopoda can give rise to both epithelial and connective tissues, but it can also generate nerve cells and fibers of different kinds.

So, in 1872, regarding the laminarity in embryonic development, Metchnikoff was cautious and spoke only of "a quite widespread phenomenon." But because the phenomenon did not have, in his opinion, any clearly established morphological characteristics, it was natural to ask how it might be identified. And, more generally, what *kind* of phenomenon was being asserted as widespread? Is there such a phenomenon at all? For Metchnikoff, the goal of embryology at this time was to establish a general affinity of different animals and to discover the plan of their organization, but not necessarily within the Darwinian construct. He was, in fact, pulled in different directions. In some sense, the very interest in defining the types of embryonic layer formation and their future role must be considered as due to von Baer's influence (hardly Darwinian!), who argued that the earliest stages of individual development possessed the clearest phyletic information and the most authentic characteristic of their respective type *(Bauplan)*. Later stages, while reflecting phylogeny, were less reliable as evidence of its true course. But if Metchnikoff's general theoretical position of that time was von Baerian, his practical results and the conclusions obtained from them can hardly be considered as strongly supporting that position. Metchnikoff was caught in the key tug-of-war of the period.

Von Baer was a strong opponent of the recapitulationist concept of a repetition in embryonic developmental stages of lower ancestral adult forms. According to von Baer, the *Bauplan* was already established from the initiation of embryonic development and was the "mechanism" of its elaboration. Different species within a given type were then formed by individualization of this general plan, and development proceeded from the general to the individual, from a more homogeneous state to a more heterogeneous one. This common tendency created the illusion of recapitulation to that degree in which two different species proceeded in their development through similar embryonic stages. So the similarity that created the opportunity to refer to recapitulation was explained, not as the repetition (in embryonic stages) of lower adult organisms (or their features), but as a similarity between embryonic stages of different animal groups. Comparing, on the one hand, von Baer's interpretation of similarity in embryonic stages and, on the other hand, the recapitulationist idea, Stephen Jay Gould contrasted their two attitudes toward Darwinian evolutionary theory:

> There existed in 1859, two major interpretations for the significance of embryonic stages. Each had been formulated under creationist tenets, but each could be easily restructured in evolutionary guise. These were, of course, von Baer's principle that development proceeds inexorably from the general to the special and the recapitulationists' claim that embryonic stages represent adult forms of "lower" creatures. Both were quickly given their evolutionary meaning: Darwin accepted von Baer's principle but stood the original explanation on its head. F. Mueller, Haeckel, Cope and Hyatt independently recognized the irresistible promise of recapitulation as a key to the reconstruction of phylogeny. (1977, pp. 69–70)

What in fact is at stake as to whether embryonic structures are similar to ancestral embryonic or adult forms? "If the goal of evolutionary theory is only to set up a series of programmatic guidelines for the reconstruction of evolutionary trees, then it makes no difference" (Ibid., p. 73). But the stakes are established when the other aspect of the question is considered: In what way can the phyletic information be reproduced in individual development, and which principles or mechanisms provide the preservation of phyletic information?

Metchnikoff's position in the early years of his research (mid-1860s through the early 1870s) is quite clear:

(1) Comparative embryology could be applied as a tool in establishing systematic relations, which, mainly due to Darwin's influence, were regarded as a

representation of evolution. But, according to Metchnikoff, this is only an applied function of comparative embryology, which has its own tasks, independent from that of reconstructing the evolutionary tree.

(2) The independence of comparative embryology from reconstructing genealogies must be maintained as illustrated by cases of apparent disruption of any parallelism between systematic relations and the order of individual development. In those cases, it is evident that the history of a species must be understood not only as an element that should be reduced to the systematic, but also as a creative process acting "within" individual development.

(3) The disruption in the parallelism between comparative embryology and taxonomy Metchnikoff regarded as favoring recapitulation against von Baer's understanding of embryological stages. The "Darwinians" (above all, Mueller and Haeckel) insisted that the parallelism had but few and insignificant exceptions for each particular case. Not projecting the implications of the recapitulation idea, Darwinism offered little to Mueller's (and his followers') understanding. In their interpretation, recapitulation appeared in the guise of the old *naturphilosophical* idea of parallelism, and in their practice the "Darwinians" were forced to invent ad hoc hypotheses, unsupported by embryological facts.

(4) Metchnikoff's final point: Although many facts supported the recapitulationist position (as opposed to von Baer's position), von Baer was correct in determining the reality of a "type" as a mode, or "plan," of individual development, and in stating that the "plan", rather than a taxonomic position, represents the integrity of an organism. But if the *Bauplan* is to be understood simply as a succession of morphological structures, the limits of von Baer's understanding of individual development have to be overcome. To comprehend the basic plan, an activity (or activities?) running through individual development must be found for integration, otherwise no plan will be discovered and the opportunity to define the integrity of organisms will be lost.

The picture was obviously incomplete; the data conflicted and no coherent pattern emerged. Metchnikoff concluded that his own attempts (as well as those of others) to comprehend the embryonic layers using only morphological and histologic criteria in invertebrate development had failed.

After 1872, Metchnikoff radically changed the intellectual goals and methods of his embryological studies. Correspondingly, he altered the range of his scientific objectives. If in the previous period he had regarded any problem of phylogenic reconstruction as, at best, an application that contributed little to the pursuit of comparative embryology's proper goals, in the later period he considered metazoan phylogeny (and especially the problem of multicellular origin) as the decisive aspect of comparative embryological studies. He regarded the approach to embryology through phylogeny, not merely as a morphological problem, but rather as having physiological significance. Earlier, he had studied a broad variety of animal groups. Now (in accordance with the central problem – the origin of metazoans) he concentrated primarily upon the lowest organisms – medusae and sponges. This shift of his general attitude towards comparative embryological studies might be regarded as a logical development of his scientific research from the earlier period, but we see the tortuous path by which he finally arrived at his later interests. In a sense, he is vividly illustrative of the shifting foci of biology in this period, as the key theoretical foundations of embryology shifted from a von Baerian essentialism to a fully evolutionary orientation.

Metchnikoff's most important conclusion from his research of the embryonic layers was pessimism concerning the possibility of establishing their universal morphological and histological identity. This pessimism marked the first failure to identify objective principles that would provide a general approach to comparative embryological studies. At least in this case, the sovereignty of comparative embryology failed to achieve its desirable objective confirmation. Thus, Metchnikoff's groping for some universal principles that governed development was the first step towards his general theoretical shift. The second step in determining Metchnikoff's new position was the recognition of alternative patterns forming the second embryonic layer (i.e., not invagination, but delamination or introgression). Henceforth, his interest was directed to the problem of mesoderm separation from the endoderm, and to a major preoccupation with the development and role of mesodermic cells. The third factor contributing to Metchnikoff's mature theoretical posture was Haeckel's gastraea theory, which appeared in the second volume of *Die Kalkschwaemme* in 1872. In the famous polemics against that theory, Metchnikoff elaborated a new vision of the relation between embryological and genealogical research to guide his own studies. He moved from an implicit intellectual possibility, to an explicitly argued theoretical position, and in the process finally abandoned a von Baerian orientation for a Darwinian one.

From 1873 to 1878, Metchnikoff published a series of papers on sponges, where he elaborated his opposition to Haeckel's "philosophical scientific method," developed his criticism of the gastraea theory, shifted his own attention to the origin and the functions of the mesodermic cells, turned to the problem of intracellular digestion, formulated his physiological approach to the task of genealogical reconstruction, and finally proposed his famous "parenchymella" hypothesis. Another avenue of research devoted to the development of the planarians (started in 1877 and continuing through the mid-1880s) was subjected to the same goals and to the same scientific logic. Thus, all the major steps leading to his major insight, including formulation of the parenchymella hypothesis, had been taken five years prior (around 1878) to his presentation of the phagocytosis hypothesis (1883), which was to fully articulate an evolutionary theory first sought in his embryological research. During this period, the polemic with Haeckel played an exceedingly important role, at times even appearing to be the very source of Metchnikoff's theoretical inspiration (discussed in detail in Tauber and Chernyak, 1991). Earlier, as noted, he had expressed his scepticism in regard to Mueller's and Haeckel's belief in a clear ontogenic-phylogenic parallelism, considering the respective tasks of phylogenic reconstruction and comparative embryology largely independent endeavors. With little at stake, their intersection was tangential. But under a different rubric, immunology, Metchnikoff would again attempt to erect a grand unifying theory, one built on universal evolutionary principles. His intellectual optimism, and creativity, never waned.

Alfred I. Tauber

NOTES

[1] Thus the immunology studies began in the early 1880s, shortly after Louis Pasteur and Robert Koch had established the bacterial etiology of infectious diseases. The medical importance of these discoveries no doubt spurred Metchnikoff to pursue his own studies of host defense, which he regarded as the province of phagocytic cells. These wandering amoebocytes, as noted, had previously been examined as part of Metchnikoff's embryological investigations, which were highly sophisticated and central to the then current debates concerning evolution. The analysis of this foundation in the context of his later work has been detailed in *Metchnikoff and the Origins of Immunology* by Alfred I. Tauber and Leon Chernyak (see ref., 1991) and summarized in Tauber (1991).
[2] Metchnikoff wrote in 1869, apparently not without some patriotic sentiments: 'The first scientific basis of embryology, that is the history of animals' development, was founded by the former St. Petersburg academicians, Pander and Baer" (Metchnikoff, 1869). Von Baer, Metchnikoff asserts here, was the true founder of comparative embryology.
[3] Jane M. Oppenheimer writes that Huxley, as early as 1849, "had appreciated the fundamental relationship between the body-layers of invertebrates and the embryonic layers of vertebrates," but in Huxley's work of 1869 *(An Introduction to the Classification of Animals)* he does not mention any relation between coelenterate ectoderm and endoderm on the one hand, and embryonic serous and mucous layers on the other; Oppenheimer concludes: "Yet twenty years later he and all other investigators were still waiting to utilize the generalization in any way, even for pedagogic reasons" (Oppenheimer, 1967). As a generalization, the idea was well suited for its time and need not have awaited the Darwinian era for its utilization. The conclusion is confirmed by the support that Zaddach received from Huxley in the 1850s and 1860s. It seems as if Huxley's modesty in 1869 (in respect to the idea that he first expressed in 1849) is most easily explained by the studies of August Weismann and Metchnikoff, begun in 1864–65, which created a skepticism in respect to (1) the natural-philosophical form of the idea drawing a parallelism between ontogeny and phylogeny, and (2) the old intention to order, in a unilinear pattern, the animal realm.
[4] In one of his first papers (1866) on insect embryology, Metchnikoff noted how profoundly the research of the 1850s had been influenced by Etienne Geoffroy Sainte-Hilaire's teaching: "In those times they tried to follow, as strictly as possible, the analogy with typical vertebrate development. The difference in the positions of the yolk, the difference in the positions of the nervous system, they tried to explain by the correspondence of the back side of Vertebrata to the abdominal side of Arthropoda."
[5] See, Chernyak and Tauber (1990) and Tauber and Chernyak (1991) for a detailed discussion of Metchnikoff's reconstructive history.
[6] Metchnikoff, 1866d, pp. 391–392. However, Metchnikoff's first research in embryology was immediately provoked by Nikolay Wagner's discovery of the phenomenon that was later defined by von Baer as "pedogenesis" (parthenogenetic reproduction by insect larvae structurally unable to copulate). (See. E. Metchnikoff, 1865a. See also von Baer's article, 1866, and Metchnikoff's letter to von Baer [Metchnikoff, E. 1866a].) Note that Metchnikoff began his embryological study with the phenomenon that later played a role in undermining the strict parallelism between ontogeny and phylogeny.
[7] After Olga Metchnikoff's book (1919, 1921), this story dominated Russian literature (Dogel and Gaisinovich). A similar interpretation was made by Metchnikoff's protégée, Besredka (Besredka, 1979). Daniel P. Todes, in his comprehensive chapter on Metchnikoff (Todes, 1989), repeats the same version of Metchnikoff's inspiration as being due to the evolutionary hypothesis and Fritz Mueller's ideas.Only by following Metchnikoff's development from his earliest work and by carefully reconstructing his concept of "disharmony" are we able to recognize his retrospective rationalization. (see, Tauber and Chernyak, 1991)
[8] As already noted, before the late 1860s, fixed microsections had not yet been developed, and in many cases Metchnikoff could observe only living larvae. Therefore, he was limited to studying the development of external structures, and only in early development could he observe the emergence of internal organs, as far as transparency allowed. This technical limit was the principal restriction in studying the role of embryonic layers in organogenesis. In fact, the very existence of invertebrate embryonic layers was unresolved for the same reason. See A. D. Nekrasov.
[9] During this period, Metchnikoff did in fact obtain important results, and he was highly productive. Some specialists consider certain of his findings as epoch making – for example, "Studien ueber die Entwicklung der Echinodermen und Nemertinen" (Metchnikoff, E. 1869c), in which all the basic problems of echinoderm development (enterocoelom, metamerism, metamorphosis) were first correctly formulated. Metchnikoff was the first (1869c) to observe coelom development in adult animals without metamerism. He described the basic processes long before the Hertwigs' report in their *Coelometeorie* (1881). (See Nekrasov, and Belkin).

[10] Dogel and Gaisinovich state that while Weismann (in 1864) had failed in understanding the developmental stages of the embryonic layers in insects, "Metchnikoff in his studies of Hemiptera and Diptera *perfectly* [italics ours] demonstrated the differentiation of the two primary embryonic layers" (SBW, p. 694). The assertion is a strange one, and the more so because in 1886 Metchnikoff himself, in recalling the events of twenty years before, wrote quite openly: "among Arthropoda I failed to discover with sufficient clarity the embryonic layers in insects" (see E. Metchnikoff, 1886, quotation on p. 421). The two quotations are found in the same book, but the discrepancy can be explained (at least partially) by the impression created by Metchnikoff's own revised version of the events.

REFERENCES

R.I. Belkin, "Embriologicheskie issledovaniia I. I. Mechnikova v ocenke ego sovremennikov," in ACW, III, 438–479.

Besredka, A. (1971) *Histoire d'une idee: L'Oeuvre de E. Metchnikoff* [Paris: Masson, 1921; trans. Abraham Rivenson and Rolf Gestreicher, *The Story of an Idea: E. Metchnikoff's Work* [Bend, Ore.: Maverick, 1979

Chernyak, L. and Tauber, A.I. (1990) "The Idea of Immunity: Metchnikoff's Metaphysics and Science.." *J. Hist. Biol.*: 23:187–249.

Dogel and Gaisinovich. SBW, pp. 693–694

Gould, S.J. (1977) *Ontogeny and Phylogeny*, Cambridge: Harvard University Press, pp. 69–70.

Haeckel, E. (1872) *Die Kalkschwaemme: Eine Monographie.* Vol.1, Berlin:G.Reiner.

Huxley, T.H. (1858) "On the Agamatic Reproduction and Morphology of *Aphis*," *Trans. Linn. Soc.* London, *22:* 193–220, 221–236.

Kovalevsky, A.O. (1866)"Entwicklungsgeschichte der einfachen Ascidien." *Met. de L'acad.* de *St. Petersbourg.*

Leuckart, R. (1858) "Fortpflanzung und Entwicklung den Pupiparen und Beobachtung an Melophagus ovinus," *Abh. Nat. Gesell. Halle, 4:*145–226.

Metchnikoff, E. (1863) "Neskol'ko slov o sovremennoi teorii proiskhjozhdeniya vidov". Published for the first time in I.I. Metchnikoff. *Izbrannye biologocheskie proizvedeniya.* Moscow: Academy of Science of the USSR, 1950.

Metchnikoff, E. (1863) "Untersuchunsen Ueber den Stiel Otr Vorticellen." *Arch. Anat. Physiol. wiss. Med..* pp. 180–186.

Metchnikoff, E. (1864) "Nachtraegliche Betnerkungen ueber den Stiel die Vorticoelinen". *Arch. Anat. Physio. wiss. Med.* pp. 291–302.

Metchnikoff, E. (1865a) "Ueber die Entwicklung der Cecidomyenlarven aus Pseudoovum," *Arch. Naturgesch.,* 1, 304–310.

Metchnikoff, E. (1865) "Ueber die Entwicklung von *Ascaris nigrovenosa*". *Arch. Anat. Physiol. wiss, Med., 4:409–420.*

Metchnikoff, E. (1865) "'Ueber die Entwicklung der Cecidomyenlarven" *Archiv. Naturgechichle (Berlin).,* 7:304–310.

Metchnikoff, E. (1866a) "Issledovanija o dvukrylykh nasekomykh." *Zapiski akademii nauk.*(Saint-Petersburg): 10(1):78–84.(Reprinted in AC, Vol. 2, pp. 56–60.)

Metchnikoff, E. (1866b) "Untersuchungen ueber die Embryologie der Hemipteren." *Z. wiss. Zool.*: 16(l):128–132.

Metchnikoff, E. (1866c) "Ueber *Geodesmus bilineatus (Fascicola teffestris* 0. Fr. Mueller?) eine europaeische Landplanarie." *Bulletin de l'academie des sciences de St. Petersbourg:* 5:433–447.

Metchnikoff, E. (1866d) "Embryologische Studien an Insekten," *Z. wiss. Zool.*, 16, 1, pp. 389–390.

Metchnikoff, E. (1866e) "O razvitii nizshikh rakoobraznykh v yaice". *Naturalist*: 2:65–72. (Reprinted in AC, Vol. 2, p. 33).

Metchnikoff, E.(1867) Dissertation for the master's degree in zoology, St. Petersburg. (reprinted in ACW, 11, 145–177).

Metchnikoff, E. (1868) Dissertation for the doctorate in zoology, St. Petersburg, in *Zapiski Akad. Nauk, 13,* 1, Appendix 1, pp. 1–48. (Reprinted in ACW, II, 17 8–207).

Metchnikoff, E. (1869a) "Sovremennoe sostoyanie nauki o razvitii zhivotnykh". *Zhurnal Minister'stva narodnogo* prosvescheniya.: 158–186, March 1869. (Reprinted in AC, Vol. 2:254–276.)

Metchnikoff, E. (1869b) "Entwicklungsgeschichtliche Beitraege." *Bulileten akademij nauk.*: 13:709–732, sec. "Melages biologiques". (Submitted to the joumal in 1868.) (Reprinted in AC, Vol. 2. pp. 239–250.)

Metchnikoff, E. (1869c) "Studien ueber die Entwicklung der Echinodermen und Nemertinen". *Mem. Akad. Nauk,* ser. 7, 14, 8 1869, 1–73.

Metchnikoff, E. (1871) "Embryologie des Scorpions," *Z. Wiss. Zool.*, 21 (1871), 204–232 (submitted to the journal in 1869; published as a separate book in 1870, Leipzig).

Metchnikoff, E. (1872) "Zur Entwicklungsgeschichte der einfacher Ascidien." *Z. wiss. Zool.* 12:339–347

Metchnikoff, E. (1886) *Embryologische Studien an Medusen: Ein Beitrag zur Genealogie der Primitivorgane* Vienna; reprinted in Russian, SBW, pp. 271–472.

Metchnikoff, E. (1893) *Lectures on the Comparative Pathology of Inflammation.* English translation by F. A. Starling and E.H. Starfing. London: Kegan Paul, Trench, Triibner, 1893. Reprinted by Dover Publication, New York, 1968.

Metchnikoff, E. (1901, 1905) *Immunity in Infective Diseases.* English translation by F.G. Binnie. Cambridge: Cambridge University Press and New York: Macmillan, 1905. Reprinted by Johnson Reprint Corp., New York, 1968.

Metchnikoff, Olga (1921) *Life of Elie Metchnikoff.* Trans. E.R. Lankester; London: Constable, 1921.

Mueller, F. (1869) *Fuer Darwin.* Leipzig: W. Engelmann. English translation by W.S. Dallas, *Facts and Arguments for Darwin.* London: John Murray.

Nekrasov, A.D. "Raboty 1. I. Mechnikova v oblasti embryologii," in ACW, III, 401–437.

Oppenheimer, J.M. (1967) *Essays in the History of Embryology and Biology,* Cambridge, Mass. and London: MIT Press, 1967, p. 269.

Pander, C. (1817) *Beytraege zur Entwickelungsgeschichte des Huehnchen's im Eye.* Wuerzburg; and *Dissertatio inauguralis, sistens historiam metopmorphoseos, quam ovum incubatum prioribus guinue diebus subit.* Wuerzbtirg.

Tauber, A.I. (1991) "The Immunological Self: A Centenary Perspective." *Persp. in Biol. and Med.* 35: 74–86.

Tauber, A.I. and Chernyak, L (1991) *Metchnikoff and the Origins of Immunology.* New York: Oxford University Press.

Tauber, A.I. and Chernyak,. L. (1989) "Metchnikoff and a Theory of Medicine." *J. Royal Soc. Med.*: 82:699–701.

Todes, Daniel P. (1989) *Darwin without Malthus: The Struggle for Existence in Russian Evolutionary Thought* [New York and Oxford: Oxford University Press, 1989, pp. 82–103.

von Baer, K.E. (1828–1837) *Ueber Entwicklungsgeschichte der Thiere.- Beobachtung and Reflection.* Koenigsberg: Borntraeger.

von Bayer, K.E. (1866). "About Professor Wagner's Discovery of Asexual Reproduction of Larvae," *Zapiski Akad. Nauk,* 10, 1, Appendix, pp. 1–77.

Weismann, A. (1863) "Die Entwicklung der Diptera im Ei, nach Beobachtungen *an Chironomus speciosus, Musca vomitoria und Pulex canix,*" *Z. Wiss. Zool.*, 13:107–220.

Weismann, A. (1863) "Die nachembryonale Entwicklung der Musciden nach Beobachtungen an *Musca vomitoria* und *Sarcophaga carnaria,*" *Z. Wiss. Zool.,* 14:187–336.

Weismann, A. (1864) *Ueber die Entwicklung der Dipteren.*Leipzig

Zaddach, G. (1854) *Untersuchungen ueber die Entwicklung und den Bau der Gliederthiere, Vol. 1, Die Entwicklungen des Phryganiden-Eies* (Berlin).

CHAPTER 1

ON THE DEVELOPMENTAL LIFE-HISTORY OF *MYZOSTOMUM*[1] (1866)

Despite the serious studies of Loven, Semper, Schultze and Schmidt, *Myzostomum*, which parasites species of *Comatula*,[2] still belongs to those mysterious organisms to which no definite systematic position has yet been assigned. Until now, opinions on the nature of *Myzostomum* are so diverse that some researchers, Max Schulze [1] in particular, list this animal in the Trematoda, while others place it in the Arthropoda.

All agree[3] that the sequence of development gives us, first and foremost, important clues for determining the systematic position of doubtful organisms. That is why, during my stay in Naples this autumn, I paid particular attention to the development of a species of *Myzostomum*, namely *Myz. cirriferum*. As to the literature on this subject, the only work that need be mentioned is that of Semper [2].

Since (like Semper) I was unsuccessful in trying to produce artificial fertilizations of *Myzostomum*, in order to study its embryology I had to search for its eggs in the muddy sediment in a jar of *Comatula*.

Fertilized eggs are round and do not show the hardening indicated by Semper in his Fig. 6 (Plate IV). This can be found only in unfertilized eggs. Fertilized eggs have only a simple egg membrane; there is neither yolk membrane nor amnion. The contents of the egg consist partly of fine grained dark yolk, but mostly of a pinkish substance (Fig. 1 [all figures refer to Semper]).

The first embryological process is the division of the cell's contents into two. It is quite remarkable that one of the newly formed cells is always larger than the other (Fig. 2). The bigger (yolk) cell also soon divides into two (Fig. 3). [Subsequent] divisions follow so quickly, one after another, within a minute at most, that they are easily missed.

Following these initial divisions, [the smaller cell] divides into four spheres (Fig. 4), almost equal in size. Thereafter, three of these cells divide almost immediately (Fig. 5), while the division of the fourth is delayed for some time (Fig. 6). Thus gradually the entire egg mass is transformed into a multitude of small polygonal cells, with pale nuclei (Fig. 7). During subsequent development, the central part of the dividing egg is transformed into a grainy mass in which the cell divisions are lost, and which serves, it seems, as nutrient material for the outer cells.

At this stage, vibrating cilia appear on the surface of the embryo; then, after a time, the fully developed embryo ruptures the egg membrane and emerges (Fig. 8). The newly hatched *Myzostomum* larva is oval, with a length of 0.05 mm (Fig. 9), and is covered with a "shell" consisting of polygonal cells (Fig. 8A). This covering is dotted with numerous vibrating cilia, long and separate from each other, by means of which the larva slowly propels itself. One can also discern a very thin, surface cuticle, with connections to the larva; this cuticle was apparently formed in the egg.[4]

This stage in development is reached in approximately 24 hours, after which one can see some larvae swimming freely in the jars.

To my regret, in spite of many experiments, I could not keep larvae alive long enough to follow their further development step by step, but even the earliest [postlarval] stage observed by me differs from the ciliated larva in its developing organs, most obviously by [the presence of] an alimentary canal and its extremities [the mouth and anus].[5] Nevertheless, I think that the gap in developmental history is not very large, since those stages not observed by me seem to be quite transitory and would probably be revealed by further careful investigation. It can hardly be supposed that there is any possibility of significant dispersal of larvae during this stage of development, since, as already mentioned, the capacity for movement in *Myzostomum* larvae is very limited.

The youngest individuals with chaetae (Fig. 10) are already 0.12 mm long, cylindrical in shape, with a clearly-visible head (p in Fig. 10). At the [anterior] end of the head, one can see a mouth opening, which leads to a muscular, oval, reversible pharynx. The pharynx, still without papillae, leads to a very straight hind gut, which differs from that of the mature animal by the absence of branched diverticulae, and it ends in a terminal anus. Concerning the appendages seen in these larvae, with the exception of one larva, all have one pair of podia, and this one is developing a second pair. The first pair of appendages have one chaeta which is slender and hook-shaped and another which still lacks its terminal plate. The second pair of appendages will later have two pairs of slender chaetae, but these have not yet appeared. This seems to indicate that this stage of development [with two pairs of appendages] is preceded by the stage with only one pair.

The stage of development with two pairs of equally developed appendages (Fig. 11) is much more common. This stage is perhaps the most protracted, and includes individuals ranging in length from 0.12 to 0.16 mm. Other details of development do not differ from the previous stage, except that the cuticle is more prominent, which can be regarded as preparation for the molting that follows.

A form of larva resembling those just described was also observed by Semper. However, I regard his illustration as not quite accurate, since he shows the leg rudiments longer and more widely spaced than in reality. Thus his larva looks much more like a tardigrade or a tick than a *Myzostomum* (cf. Semper's Fig. 9, Plate 3 and my Fig. 11).

I observed very few larvae with three pairs of appendages, without any other special features. I never observed a stage with four pairs of appendages, so I cannot say whether it occurs, although I consider it quite possible.

The youngest specimens of *Myzostomum* with the final number of appendages (five pairs) have wide plates at the ends of the chaetae (Fig. 12), not seen previously. The most posterior pair of legs is the least developed. The unbranched intestine contains assorted food particles, consisting, it seems, of diatoms. Only later, in individuals of 0.45 mm, three protrusions of the intestine in the form of blind offshoots may be observed (Fig. 13), but as yet there is no branching of the intestine. The papillae of the reversible pharynx and cirri are also absent.

These observations clearly show that the appendages, so characteristic of this animal, develop gradually, in pairs, from anterior to posterior. This form of development already allows us to add *Myzostomum* to the Articiculata,[6] but I am firmly convinced that this animal is not an arthropod, as recently claimed by certain scientists. This erroneous opinion is based primarily on the observations of Semper, the first to give reasons for including *Myzostomum* in the Arthropoda. He claimed the following:

Points in favor of the inclusion of *Myzostomum* in the Crustacea are indicated by the larval stage of *Myz. cirriferum*, which closely resembles the larval forms of all parasitic Crustacea.[7] If one takes into consideration the great similarity of some developed animals, e.g. *Linguatula*,[8] to worms, it is easy to agree with Von-der-Heaven, who grouped them together.

Opposing these opinions of Semper, von Carus [3] speaks out even more definitely: "Judging by the larval forms, these animals (*Myzostomum*) seem to be most probably related to the Tardigrada, or to *Peltogaster*, or to *Saculina*."[9] This opinion is also shared by Carlir [4] and Leydig [5].

Semper's data on the developmental forms of *Myzostomum* was incomplete (see above), but, apart from this, his views cannot be accepted as correct because *Myzostomum* larvae with two pairs of appendages do not show any similarity to the highly characteristic nauplius larvae of many crustaceans. Their appendages are organized quite differently: they never have inner chaetae resembling those in *Myzostomum*, but they generally have pinnate setae, which never occur in *Myzostomum* larvae. In addition, we know of no crustacean nauplius with two pairs of uniramous appendages, but such a stage occurs in the development of *Myzostomum*. And, finally, all forms of nauplius have a large labrum [upper lip] and a clearly differentiated telson, not seen in larvae of *Myzostomum*. The latter, on the contrary, are distinguished by inner bristles and a well developed reversible pharynx, features that are completely absent in crustacean larvae.

All the above clearly demonstrates the untenability of this widely held opinion. To my mind, no further discussion of this question is needed to show that *Myzostomum*, with a larval stage covered by cilia and with all its anatomical and embryological peculiarities, is not a member of the Arthropoda but a representative of the diverse groups that make up the Vermes.[10]

The view, first expressed by Max Schulze, that *Myzostomum* should be placed amongst the trematodes is questionable. On the one hand, the common features of *Myzostomum* and trematodes have no significance. For instance, suckers, which are common to both groups of animals, may be seen in other groups of worms (Hirudinea, *Leucodora*,[11] some planarians, Cestoda, etc.). Further, a branched intestine is found not only in trematodes but in other groups of worms also. On the other hand, there are many factors which prevent placing *Myzostomum* within the trematodes. The appendages, typical of the former animal, the cirri and the anus have no counterparts in trematodes. And even more important differences have been revealed by their respective life-histories, simply that *Myzostomum* is shown to be an evolved, multi-organ animal (see above).

Because of these observations, it seems to me that the most natural conclusion is to regard *Myzostomum* as a parasitic member of the Annelida, and its parasitic way of life is responsible for all the peculiarities of its organization which distinguish it from any related form. I shall endeavor to substantiate my statement with more detailed descriptions.

The **skin** of *Myzostomum* is indistinguishable from that of an annelid, with its fairly well developed cuticle with groups of chaetae on the surface; such a combination is completely alien to all other forms of worms. The construction of the **alimentary system** also shows many significant similarities: the reversible pharynx of *Myzostomum*, with terminal papillae (especially in *Myz. cirriferum*) is almost identical to the same organ in many annelids (e.g. in some Hesionea and Phyllodocea). The branched intestine of *Myzostomum* merely repeats a peculiarity already familiar to us in *Aphrodite* and related forms.

The construction of the **appendages** in *Myzostomum*, which, as previously emphasized by Semper (cf. p. 52 in paper quoted above) differs drastically from that observed in the Arthropoda, definitely confirms my position, since analogous structures can be found only in the Annelida. Not only the parapodia, but also the chaetae are arranged identically in both groups (*Myzostomum* and Annelida). Very importantly, in *Myz. cirriferum* we also observe the existence of cirri, which always occur in a definite number, twice as many as the number of parapodia. These cirri show a definite relation to the parapodia, and consequently may be positively compared to the undivided cirri of many Annelida.

The absence of **blood vessels** in *Myzostomum* should perhaps be regarded as an objection to my view, but some representatives of the Annelida without blood vessels (e.g. *Aphlebina, Capitella, Glycera*) have been reported.

In order to explain the peculiar organization of the **neural system** in *Myzostomum*, one has to take into consideration the parasitic way of life of the animal and its lamelliform body arising from this way of life. The structure of the abdominal ganglia is the same as that found in many articulates with the same body form. I, like my predecessors, did not succeed in tracing the path of the branching nerves, but I believe the existence of a circumesophageal commisure to be quite possible.

While the foregoing systems of organs in *Myzostomum* and the Annelida are found to be nearly identical, such similarity can definitely not be claimed with respect to their *sexual organs*. I am here referring not to hermaphrodites (we know of several hermaphrodite annelids, like *Protula, Spirorbis, Amphicora*), but to the cloaca of *Myzostomum*, described by Semper, which has not previously been observed in any genuine annelid. Other elements of the sexual system of *Myzostomum*, however, seem to be quite comparable to those found in the Annelida. For example, both groups have no specialized testes or ovaries. Further, sperm-ducts in *Myzostomum* are comparable to the segmental organs of some Chaetopoda[12] which have only one pair of sperm-ducts (e.g. *Parthenope*). I have to mention, however, that extremely diverse structural forms and the functioning of the sexual apparatus of many annelids have not yet been sufficiently studied, making it virtually impossible to offer explanations or to draw analogies between different species in this field.

What is now left to discuss is the question of **abdominal suckers** in *Myzostomum*, whose existence is clearly linked to its parasitic way of life. That these organs are not quite alien to genuine annelids is shown by the sucker-like tail ends [of leaches].[13]

Embryology also gives us an important vantage point for comparing *Myzostomum* with the Chaetopoda with regard to the gradual development of the rudiments of appendages. This development follows a fixed sequence, which convinces us once again that *Myzostomum* is a segmented worm related to the Annelida. I should like to compare the still incompletely known metamorphosis of this animal to the metamorphosis of annelids from what Johannes Müller called the atrocha[14] stage. I personally observed this type of metamorphosis in Naples in some annelid larvae, which I consider to be larvae of *Syllis* because of the formation of their chaetae. The youngest larvae had an overall ciliary cover (only the front end was not covered with vibrating hairs), without a distinct belt of cilia, and it had only **one pair of parpodial rudiments** with fully developed complex chaetae. The most obvious distinction characterizing the second stage is a second pair of parapodial rudiments. More developed *Syllis* larvae, which already had three pairs of appendages, were not covered with cilia.

Another similarity with *Myzostomum* is the absence of ciliary "propulsive bands", and the presence of overall ciliation seems to me to be sufficient reason for listing the metamorphosis of *Myzostomum* as belonging to the atrocha type.

Until now, we have not discussed the body shape of *Myzostomum*, but here I merely mention that the clumsy body form of this animal is wholly conditioned by its way of life. Young stages are much more elongated than adults, and, in essence, differ little from annelids.

I regard the foregoing reasons to be quite sufficient to show the annelid-like nature of *Myzostomum*, and consequently I consider this animal to be a representative of a particular group of Chaetopoda, which may be called Chaetopda Ectoprasitica.

Göttingen, 25 December 1865.

REFERENCES

[1] Report of several zootomic studies carried out in the Fall of 1853 on a Mediterranean coast. Verhandlungen der Wurzburger phys.-medic. gesellschaft, Bd.IV, S.225 ff.

[2] Zur Anatomie und Entwicklunsgeschichte der Gattung Myzostonum Zeitschrift fur Wissenschaftliche Zoologie, Bd. IX, 1857, S.48 ff.

[3] Carus u. Gerstaecker, Handbuch der Zoologie, 1863, S.480.

[4] Examen sur les princ. classific. etc. 1861.

[5] Vom Bau thierischen Korper, I, 1864, S.135.

NOTES

1 Today, most zoologists follow Metchnikoff in regarding the Myzostomida (or Myzostomaria) as an order of Polychaeta. This is supported by the appendages of *Myzostomum*, which resemble polychaete parapodia and have a similar pattern of development. The fact that the ciliated early larva of *Myzostomum* shows some resemblance to the larvae of some polychaetes is, in our opinion, of little importance. Several distantly related groups have similar larvae, and perhaps the "larva" which shows the greatest resemblance to that of *Myzostomum* is the zoospore of the green alga *Debresia marina*, which is not only of similar shape but also lacks an alimentary canal.

2 A crinoid echinoderm. We have followed the recommendation of the International Code of Zoological Nomenclature that generic and specific names (only) should by placed in italics. The Latin names of phyla, classes, orders, families, etc., should have an initial capital (e.g. Arthropoda, Crustacea) but not their English equivalents (e.g. arthropods, crustaceans).

3 This was true in 1865 and is still generally accepted. Today, however, there is also a Williamson heretical view that types of development can be transferred between different groups. (Donald Williamson. *Larvae and Evolution. Towards a New Zoology*, New York: Rutledge, Chapman and Hall, 1992)

[4] Some later workers argue that it is the original egg membrane.

[5] The ciliated larva of *Myzostomum* lacks mouth, anus and alimentary canal. The origin of these organs is still unknown, and, although *Myzostomum* is usually regarded as a protostome, this is not based on direct evidence.

[6] Annelida + Arthropoda

[7] Whether Semper meant parasitized or parasitic crustaceans, in either case Metchnikoff is more likely correct inasmuch as no crustacean has a larva closely resembling that of *Myzostomum*, and, like all arthropods, crustaceans are devoid of cilia throughout the life cycle.

[8] *Linguatula* is an arachnid, which suggests that Semper tended to lump all arthropods together.

[9] *Peltogaster* and *Sacculina* are members of the Rhizocephala, parasitic on crabs and hermit-crabs. Their larvae resemble those of barnacles, but the Rhizocephala are probably not arthropods.

[10] The phylum Vermes was used to cover all segmented and unsegmented worms. It is no longer regarded as a single phylum, but zoologists argue about the number of phyla involved and their relationships.

[11] A member of the Polychaeta.

[12] Polychaeta + Oligochaeta. Today, the Oligochaeta (earthworms) are more usually grouped with the Hirudinea (leaches) in the Clitellata.

[13] Metchnikoff doesn't mention leeches here, but he did earlier.

[14] *Atrocha* Claparède & Metchnihoff, 1869, is the generic name of a polychaete, but this is its first use in this sense. Müller used it as a type of "with no rings", in contrast to the trochophore larva, which has an equatorial ring of cilia.

CHAPTER 2

ON THE EMBRYONIC DEVELOPMENT OF LOWER CRUSTACEA (1866)

The development of invertebrate animals has been the primary object of research for naturalists for the past 10–15 years and remains so. The majority of the best zoologists of our time have come from this field, and, indeed, each successive step in the development of lower animals reveals new phenomena and significantly broadens our horizons.

Much has been done in this field of science, but a great deal of knowledge is still missing. What is lacking is precisely that information which might permit a comparison of the development of different animals and a generalization of the laws of development. In other words, what has not yet been developed is comparative biology. It is common knowledge, however, that scientific facts acquire their full significance only when correlated with a series of related phenomena, *i.e.* by comparison.

Having no hope of remedying this situation immediately, I have nevertheless begun to study subjects which might yield significant results for comparative embryology. With this goal in mind, I have studied many invertebrate animals and collected much material, which I intend to report in a series of papers. Here I should like to draw the reader's attention only to the phenomena of embryonic development (*i.e.* development within the egg) of certain lower Crustacea, namely the genus *Balanus* (from the curl-footed Cirripedia), the genus *Sacculina* (parasites closely related to the Cirripedia), and to the embryology of a genus normally found in fresh water, *Cyclops* (from the so-called Copepoda). It is well known that these genera belong to closely related groups,[1] and precisely because of this, a comparative study of their embryology seems pertinent.

The larvae that emerge from the eggs of these animals have basically the same characteristics: a conical shape and locomotion by three pairs of legs, the first two of which correspond to the [two pairs of] antennae and the last to the mandibles of the adult animal.[2] In addition, larvae of these animals have an unpaired, simple eye, and a large lobe on the ventral side of the head, the so-called labrum, with the mouth located at the end.[3] Such larvae, which share the name *Nauplius*[4], [1] have considerable swimming powers, which accounts for their appearance mostly at the surface of fresh and marine waters. As they grow, [the larval features] fade, and at each molt they acquire new features bringing them ever closer to the adult form.[5]

We shall not follow all these larval transformations but instead confine ourselves to processes that take place in the egg, and which result, in the three cases studied, in the larvae described above.

The first step in the development of our crustaceans is the separation of the contents of the egg into several parts – so-called differentiation.[6] Differences in the development of the [three] genera studied are already quite evident at this stage. In the egg of *Balanus*, two parts, distinguished by their content, become separated first:

the larger consists of a conglomerate of relatively large yolk grains, while the smaller seems to the observer to be formed of colorless matter (so-called protoplasm) with a considerable number of very small grains. I should like to point out here that the first part of the dissociated egg forms *nutritive* yolk, the second part *formative* yolk. The actual process of division into cells occurs only in the latter part; tissues and organs of the body are formed from cells derived from the formative yolk. Ontogeny is completed according to rules common to all animals, with such a multiplication of embryonic cells resulting in the formation of an embryonic membrane. This membrane consists of only one layer of cells and envelopes the inactive nutritive yolk. This nutritive yolk is in contact with all the cells of the membrane and provides them with food, thereby permitting growth and the ability to develop to maturity. Multiplication of these cells occurs predominantly on the side corresponding to the ventral part of the larva; thus a relatively thick cellular layer forms here (the so-called *primary zone*), and, as a result, the nutritive yolk becomes so thin that it contacts [only] the outer membrane of the egg.

I have already noted that in both *Sacculina* and *Cyclops* the process of differentiation and formation of the initial parts of the future embryo occurs differently from that just described for *Balanus*. The former animals show no early separation into nutritive and formative yolks; in *Sacculina* such a division occurs much later, for the initial divisions of the contents of the egg are first into two equal parts, then into four. Thus *Balanus* differs in that only the formative yolk replicates itself by division, whereas in *Sacculina* the entire content of the egg is involved in this process (at least initially). According to accepted terminology, the first case is characterized by *incomplete cleavage of the yolk*, the second by *complete cleavage*.

Some naturalists, like Philippi and F. Müller, [2] still believe that cirripedes have complete cleavage, but they are mistaken. The two parts just described, namely the nutritive and formative yolks, cannot be compared to two dividing spheres.

In the egg of *Sacculina*, rapid differentiation of the formative part is seen only after division of the yolk into four, and consists of four small cells filled with very small granules. These cells continue to multiply by division to form an embryonic membrane, while the four large spheres which form the nutritive yolk lose the capacity to divide and remain intact until later stages of development, when they dissolve, as it were, into numerous fragments.

I shall not describe the subsequent changes in the egg of *Sacculina*, since they differ from the development of *Balanus* in minor details only, and such details are more appropriate to a specialized technical paper. Here I should like to concentrate only on the aforementioned peculiarities.

There is no doubt from what has already been said that both *Balanus* and *Sacculina* both have nutritive yolk. At the same time, however, it is evident that in the first animal this yolk separates from the formative yolk before the process of cleavage begins, while in *Sacculina* this separation takes place later, just after the first two cleavage divisions. This fact provides the ultimate refutation of the opinion, still widespread, that nutritive yolk is found only in eggs with incomplete cleavage. Although this opinion has been questioned, it is not yet regarded as rejected. Proponents of this theory attempt to prove that there is no solid division between complete and incomplete cleavage, nor, for that matter, a firm division between nutritive and formative yolk, and, on this basis, they erroneously claim that nutritive yolk occurs only in eggs with incomplete cleavage.

I am quite convinced, on the basis of my data, that the opinion that nutritive yolk occurs only in eggs with incomplete cleavage is incorrect, and I cannot agree with those who claim that there is no significant difference between nutritive and formative yolk or between complete and incomplete cleavage. Nutritive yolk differs from the formative type both in composition and in function. The former usually consists of large yolk grains and has no cellular structure, while formative yolk is in the form of cells, each of which, in addition to the nucleus, is made up of colorless protoplasm with very small particles in it. The crucial point is that the formative yolk becomes detached from the nutritive yolk at different times in different animals. For example, in *Balanus* the formative yolk becomes separated even before cleavage; in this case we see only incomplete cleavage, affecting nutritive yolk only. In other animals (*e.g. Sacculina*) the first phase of cleavage precedes differentiation of the two types of yolk. We must conclude that although *nutritive and formative yolk differ drastically from each other, the mere presence of nutritive yolk is not linked with either type of cleavage.*

In other animals we see that the two types of yolk separate at entirely different stages of development. In marine planarians, I know of genera in which formative yolk is distinctly separated from the nutritive type while the egg is still in the ovary.

Although the occurrence of nutritive yolk is not limited to the few animals with incomplete cleavage, it still cannot be claimed (as some naturalists do) that all animals with complete cleavage have nutritive yolk. On the contrary, animals with incomplete cleavage but with nutritive yolk have been known for a long time. Such is the third genus of those considered here, namely the freshwater *Cyclops*. Eggs of this animal undergo complete cleavage, *i.e.* all the yolk content is divided into a fairly considerable number of cells. Nutritive yolk, as I have just mentioned, is not formed in the eggs of *Cyclops*, but instead we find a formation not present in either *Balanus* or *Sacculina.* I refer here to the peculiar cavity situated precisely in the middle of the egg and surrounded by a layer of newly developed embryonic cells, corresponding to the "growth cavity" (blastocoel) found originally by K.M.Baer in the frog, and thus named the Baerian cavity. Indeed the central location of this cavity in *Cyclops* uniquely shows that it occupies precisely the same position as the nutritive yolk in the other animals considered (*Balanus* and *Sacculina*). This leads to a simple law: a Baerian cavity can develop only in animals without nutritive yolk.

Let us now further analyze the embryonic development of *Cyclops*. The embryonic membrane which surrounds the Baerian cavity consists of one layer of cells which remain unchanged for a relatively long time. *Cyclops* (as Klauss discovered) presents a remarkable difference from the development of *Balanus* and *Sacculina* in the complete absence of a primary embryonic zone,[7] usually situated on the ventral surface. This fact is too significant to omit. Indeed it is well known that the embryonic zone, so important for development since the ventral nervous chain originates from it, occurs in almost all arthropods and even in many annelids. Of the widely recognized arthropods, only *Cyclops* and the Tardigrada [3] are completely deprived of an embryonic zone. Quite remarkably, these animals provide the only examples of arthropods without nutritive yolk but with a Baerian cavity; this obviously prompts the suggestion of a causal link between these anomalies. Indeed, it is quite possible that the embryonic zone, considered on its own, needs a special nutritious and supportive mass for its development and maintenance, and this turns out to be the nutritive yolk.

I return now to the genus *Balanus* [4] to determine the major steps in its development. Soon after the establishment of the embryonic or primary zone, we see the formation of three pairs of "knobs" on it, which later transform into three pairs of swimming legs. In addition to these developments, differentiation of other parts of organs occurs simultaneously. Posterior outgrowths are formed from a significantly thickened lower part of the embryonic zone, and here also begins the same process of transformation of the homogeneous mass of embryonic cells into new parts. Inner cells completely filled with small yolk granules separate themselves from other cells and form the intestinal membrane, the anterior end of which touches the nutritive yolk mass. Peripheral cells, which differ from those mentioned above by their lack of color, simultaneously form the cuticle and muscle fibers. Between these layers of tissue is an empty space which now becomes part of the body cavity of the organism. When these processes are completed, differentiation of the alimentary canal and the so-called *animal* layer of the embryo begins, starting generally from the ventral side (where the embryonic zone is situated) and spreading from there, by cell multiplication, to the dorsal region. The animal layer, [5] in addition to skin and muscle, also gives rise to the neural system, which, in our larvae, consists of one paired cerebral ganglion. The intestinal membrane now encloses the entire mass of nutritive yolk, the quantity of which gradually decreases from now on. Thus the developing digestive system does not yet have either anterior or posterior parts, *i.e.* neither mouth and oesophagus nor anus and rectum. These parts are formed in a completely different manner: invagination of the front and rear ends of the embryo results in a canal connecting the three newly developed parts of the intestine.

This is the way in which organs are developed in *Balanus*. The same method of organ development occurs in *Sacculina* and in many [other] known animals, and it may well be a common method of development in the annelids.

I have deliberately described these processes in some detail in order to facilitate comparison with the development of *Cyclops*. As already noted, these animals illustrate a remarkable feature in possessing a *cleavage* or *Baerian* cavity. At the start of its development, this cavity grows dramatically in volume and reaches its maximum at the time when separate zones appear in the body of the embryo (which still consists of a single layer of cells). Thereafter the cavity becomes smaller and smaller through the rapid multiplication of the surrounding cells. Very often it virtually disappears, although its existence should by no means be doubted. What is taking place during this period of cellular multiplication is the formation of both the animal layer and the intestinal membrane; all the rest is similar to the development of *Balanus*. The principal difference is the intestinal membrane of *Cyclops*, instead of surrounding the nutritive yolk as in *Balanus* , forms around the blastocoel, which, therefore, becomes part of the intestinal cavity.[8] The body cavity in *Cyclops* is formed in exactly the same way as in *Balanus*.

Let us summarize the major findings.

1. The body cavity in the Crustacea (as well as in other arthropods) arises as a result of the development of layers of embryonic tissue, and, of those developed, expressly between the animal layer and the intestinal membrane.
2. The intestinal canal (*i.e.* the entire alimentary canal except the oesophagus and anus) is formed in *Balanus* and *Sacculina* through absorption of the nutritive yolk, which is surrounded by the intestinal wall, while in *Cyclops* it originates from the blastocoel.

If we recall the similar topographical location of the nutritive yolk in one group of animals and the blastocoel in the other group, also noting that this cavity is found only in animals without nutritive yolk, then the identity of the intestine and its significance for both methods of formation becomes quite clear. We can thus say that *the blastocoel is always completely filled with the equivalent of nutritive yolk.*

The certainty and simplicity of such a conclusion, resulting from the similarity of the objects analyzed and their ready and direct comparability, provides a key to some generalizations, the aim of which is to understand isolated facts and to reveal the rules that govern them.

The actual formation of the cavity of the alimentary canal from the blastocoel was first spotted by Gegenbauer in the unusual worm *Sagitta* [6]. Soon after, the same phenomenon was observed by Semper [7] in *Eucharis multicornis*, an animal of the order Ctenophora. But since these facts were associated with animals with very simple and, at the same time, very distinctive organization, their broader interpretation could not be appreciated, and Gegenbauer was quite amazed by his discovery of how the intestinal canal developed. Now, however, when we compare the method of development in *Cyclops* with that in other animals, the above mentioned facts can easily be reconciled and reduced to a general rule. Indeed, in both *Sagitta* and *Eucharis* nutritive yolk is completely absent, therefore [these animals] should have a blastocoel which should develop into the intestinal cavity. Other studies now demonstrate that a similar process takes place in many animals without nutritive yolk. Thus, for example, it can be inferred from the works of Leydig [8] and Kaufmann [9] that *Paludina*[9] and the Tardigrada do not produce nutritive yolk but a blastocoel which later develops into the intestinal cavity.

A very general question automatically presents itself, namely, does every animal with a blastocoel have the same method of forming the intestinal cavity? To answer, one has firstly to review the current scientific data on the subject.

Apart from the above-mentioned types of animals, the actual existence of a blastocoel has been described in the eggs of the frog (where it was first discovered by von Baer), in the eggs of the lamprey (according to M. Schultze), and, finally, in the eggs of *Amphioxus* (as pointed out by Kovalevsky).

It is well known that, at a certain stage in the cleavage of the frog egg, a small cavity appears and grows bigger and bigger; its growth ends, however, with the appearance of a new cavity, the cavity of the so-called archenteron. From now on the blastocoel becomes smaller and smaller and finally can no longer be seen. This observation is usually interpreted as marking the complete disappearance of this cavity. Not daring to reject this opinion entirely, I would suggest rather that the actual existence of the cavity should not be denied, but it disappears from the eye of the observer for some time. The cavity is unnoticed because of the convergence of the walls that surround it, but nevertheless it still retains the capacity to expand when the walls begin moving apart. Quite possibly this process takes place in the blastocoel of the frog, *i.e.* the cavity does not disappear but is merely squeezed by the mass of surrounding cells and then lengthens into the cavity of part of the alimentary canal. Of course, this suggestion must be confirmed by experimental observations,[10] but since we do not yet have any, we should use related data. A compelling reason for this opinion is that *the blastocoel, while shrinking, is situated in that part of the egg from which the intestinal wall will be formed, i.e. in the so-called "intestino-glandular layer".* [10] This observation is sufficient to disprove the recent statement of Kovalevsky that the blastocoel changes into the general abdominal cavity of the body [or coelom]. As we have just seen, the position of the blastocoel corresponds to that of the alimentary canal rather than to that of the coelom, which, as is well known, *is always formed between the so-called animal and vegetal layers.* Such development of the body cavity has been observed in the frog by Remak; his description is quite adequate [11].

My view on the fate of the Baerian cavity, in addition to the facts just stated, is also supported (but not explained) by Remak in his discussion on the diverse origins of the different parts of the alimentary canal of the frog. Links between the Baerian cavity and the archenteron, found by the same naturalist, also support my views.

All the above comments on the frog can certainly be applied to the lamprey, *Petromyzon planeri*, as far as is known from the as yet incomplete studies of Max Schultze. The formation and position of the blastocoel [in the lamprey] is the same [as in the frog], so it cannot be suggested that it transforms into the general body cavity. From the figures drawn by Max Schulze, it is however quite possible that the cavity of the archenteron in the lamprey fuses with that of the Baerian cavity.

It follows from the excellent work of Kovalevsky that the lower vertebrate *Amphioxus* has a blastocoel which could regenerate as the general body cavity. Since this view is completely at variance with what we see in other animals, I find it necessary to discuss the proposal in detail.

Kovalevsky formed the above-mentioned opinion from observations which I consider to be rather inadequate in this particular case. In Figure 9 he shows an early embryo with a relatively large blastocoel; then he describes the next stage (Fig. 12) in which this cavity has become compressed due to deepening of a layer of cells. Kovalevsky is in no doubt about either the existence of the depression or the narrowing of the blastocoel and its transformation into the coelom. Nevertheless, a significant period of time, namely four hours, sufficient for quite dramatic changes, went unnoticed between these changes. Kovalevsky's opinion cannot, therefore, be regarded as conclusive. There is even an observation which suggests that his opinion is mistaken. In the same Figure 9, we see that the blastocoel is bounded on each side *by not one but two layers of cells*, while the coelom (Figs. 12 and 13) is surrounded on every side *by a single layer of cells*. This occurrence forces us to conclude that the topographical locations of the cavities are not identical, although Kovalevsky regarded them as the same.

An analogy reinforces this opinion. Kovalevsky's Figures 12 and 13 are indeed so similar to Gegenbauer's Figures 8 and 9 (depicting stages in the development of *Sagitta*) and to Semper's Figure 7 (where an embryo of *Eacharis* is reproduced) that it is difficult to see any difference between them. Just as we pointed out above that *Sagitta* and *Eucharis* serve as classical evidence of the transformation of the blastocoel into the intestine, why not assume that the same transformation takes place in *Amphioxus*? I think, from the analogy and observations mentioned above, that [this assumption] is almost unquestionable, and, therefore, the development of the blastocoel into the intestine cannot be accepted.

It is indeed odd that Kovalevsky himself draws attention to the great similarity between *Sagitta* and *Amphioxus* embryos, and still, in spite of this, has no doubt of the correctness of his opinion.

Kovalevsky considers that the general body cavity (coelom) is produced by the blastocoel and claims that the alimentary canal is formed by thickening (*Einstulpung*). He extends this [opinion] to the lamprey, but where are the supporting facts? In all animals studied, the alimentary canal *is never formed by such thickening but always by forming walls next to the nutritive yolk or around the blastocoel.*

It follows from the above that the formation of the alimentary canal in frogs, lampreys and *Amphioxus* seems to fit a general rule which I discovered while studying the development of Crustacea. If other observations confirm this conclusion, it will mark another contribution to the comparative method.

REFERENCES

[1] Described under this name in the previous century by Otto Müller, who considered them to be mature animals.

[2] *Archivio per la Zoologia*, 1861, and in *Moleschotts Untersuchungen etc.*, Y. IX.

[3] As is clear from the precise drawings of Kauffmann (*Zeitschrift fur Wissenschaftliche Zoologie*, Bd. 3, 1851, Tabl.IV), tardigrades do not have an embryonic primordium, and what is being formed in the center of the egg is the Baerian cavity.

[4] Fritz Müller rejects its inclusion in the Cirripidia, but [the form of development] shows, without doubt, why he is mistaken.[11]

[5] I have to say that this animal layer by no means corresponds to that described by de Philippi (op. cit.), who erroneously regarded two areas of blastoderm cells as two primary layers.

[6] Gegenbauer, *Ueber Entwicklung der Sagitta*, Halb., 1857.

[7] In *Zitschrift fur wissenchaftliche Zoologie*, Bd. IX, S.234.

[8] Loc. cit., Bd.II, S.125.

[9] Loc. cit.

[10] This is the most evident in Schtrikker's text (Zeitschrift f. wissenschaftliche Zoologie, Bd.XI) figure 5 F (mash. XXVI).

[11] Remak, *Untersuchungen uber die Entwicklung der Wirbelthiere,* 1855, S.159.

NOTES

1 These groups are said to be related because of larval similarities, but Williamson argues that the larvae were later additions (Williamson & Rice, 1996. Larval evolution in the Crustacea. *Crustaceana* **69**: 267–287.). An adult *Sacculina* is not a crustacean, but its larvae resemble those of cirripedes.

[2] Williamson challenges the supposition that the appendages of nauplii "correspond" to those of adult crustaceans, arguing that ancestors of several groups of crustaceans acquired nauplius larvae by hybridizing with members of the Nauplioida, a group of non-crustacean arthropods (see, Williamson and Rice, op. cit.).

[3] Metchnikoff was mistaken in placing the mouth at the end of the labrum. The labrum covers the mouth.

[4] We presume Metchnikoff treated *Nauplius* as a generic name, as it originally was.

[5] Not an observation, merely a guess. It is far removed from reality.

[6] for "segmentation". There is no metameric segmentation at this stage.

[7] embryonic primordium.

[8] Now considered an incorrect conclusion.

[9] A viviparous gastropod that develops by deuterostomy and enterocoely. All other known mollusks develop by protostomy and schizocoely. Williamson regards it as an example of embryonic transfer (Williamson, 1992, op. cit n. 3, p. 26).

[10] Not confirmed by subsequent observations.

[11] Darwin (*Origin of Species*) said much the same.

CHAPTER 3

ANTHROPOLOGY AND DARWINISM (1875)

I. DEVELOPMENT OF DARWINISM. – NEED FOR INDUCTIVE VERIFICATION. – SIGNIFICANCE FOR DARWINISM OF ANTHROPOLOGY: THE SCIENCE OF MAN

More than fifteen years have passed since Darwin and Wallace published the results of their studies on the origin of species and thereby laid the foundation of a new era in biological science. It is fortunate that the ideas of these two great English naturalists were not hidden "under a bushel," although this is certainly not what happens to many scientific developments. Scientists, who had long been languishing in a dearth of seminal ideas, seized the new theory with impatience and quickly brought its light to bear on the darkest areas of their own specialties. Thus, in a comparatively short time, Darwinism (as this theory has been called) penetrated comparative linguistics, political science and other schools, as well as all fields of biology. The theory has been particularly popular in Germany, where it was first made known through university curricula and public lectures and then spread to regular student seminars. Several years ago, when the Guelf University of Göttingen sought to appoint an opponent of Darwinism to the Biology Department, no such candidate could be found among the recognized biologists looking for a position, so the university had to compromise and invited a "moderate" Darwinist. The literature on Darwinism has expanded to such an extent that special bibliographies have become necessary, and even annual reports on such literature have recently started to appear. Spengel's bibliography for 1873 lists 304 authors writing on the new theory, and this catalog is far from complete.

Such rapid spread of the theory, with no pretensions to scientific proof, can be explained, on the one hand, by Darwin's extraordinary scientific intelligence, and, on the other, by the simplicity, lucidity and all-embracing completeness of the theory itself. The very concept that all organisms are connected to each other by consanguinity, i.e. that they are all descended from the same ancestors, is not at all new. It was the foundation of the Lamarckian theory, which was once adopted by many first-class scientists, and survived, in more or less the same form, until Darwin and Wallace entered the scene.[1] It is evident that one idea, no matter how powerful, is not sufficient basis for a new theory.

Darwin is quite right in mentioning in the Introduction to his book:

"It is quite conceivable, that a naturalist, reflecting on the mutual affinities of organic beings, on their embryological relations, their geographical distribution, geological succession, and other such facts, might come to the conclusion that each species had not been independently created, but had descended, like varieties, from other species. Nevertheless, such a conclusion, even if well founded, would be unsatisfactory, until it could be shown how the innumerable species inhabiting this world have been modified, so as to acquire that perfection of structure and co-adaptation which most justly excites our admiration."[2]

The entire essence of Darwinism, the theory of the origin of species through natural selection of the best-adapted organisms, can be regarded as a response to this challenge. This theory, however, should not be mistaken for the theory of general continuity of species (*Desdendennzlehre*): the former is related to the latter as a more specific concept to a more general one.

If the new theory did not apply to all organisms, its main foundations would appear to be purely deductive. For both Darwin and Wallace, natural selection is just a simple deduction from the principles of the struggle for existence. Since, as the theory claims, the entire mass of living organisms cannot survive without constraint, competition between them is unavoidable. Only those creatures which, by chance, possess certain individual characteristics win this perpetual struggle. These favorable attributes, precisely because they are so significant in the struggle, come to the fore and accumulate, progressing from mere individual peculiarities to widespread features.

Many secondary corroborative deductions have been brought forward to confirm each statement of the theory, several with supporting data, but, even today, the theory of natural selection remains virtually without any empirical confirmation. This, nevertheless, is necessary and quite obtainable. No one would reasonably demand that the process of transformation of species through natural selection be completely demonstrated in a short period of time, but we can carry out separate strides of both processes: the transformation of organisms and the struggle between organisms. Experts can discern subtle qualities that would generally be regarded as insignificant. Darwin himself mentions authorities who can tell the difference between strands of sheep's wool by infinitesimal differences in thickness, and they are thus able to solve an essential problem in artificial selection. Why do not naturalists use similar techniques in looking for indications of individual subdivisions resulting from the struggle for existence? Such cases may be encountered everywhere in nature. One should consider, for example, the enormously rapid reproduction of organisms in a limited space, like water plants in closed reservoirs or organisms which rapidly populate small islands. Who has not seen how quickly the surface of stagnant water can become covered with a thick layer of duckweed so that not a drop of water can be seen through it? This seems to be an extreme case of overpopulation and therefore must involve quite an intense struggle for existence. [1]

There is, therefore, an evident need for an inductive examination of the major ideas behind the theory, especially as the biological phenomena are so complex. It is unreasonable to expect quick results from studies covering most aspects of the theory; nevertheless we can achieve some results almost immediately by transplanting Darwinian theory to the soil of the science of man.

Anthropology deals with one species only, or, alternatively, with several closely related ones. It makes no difference to us today whether the so-called human races are considered to be varieties of the same species or separate species; the question is of no scientific significance. It is quite sufficient to know that human "races", so different in appearance, are in fact so similar to each other that they can produce descendants, and that this process includes a number of transitional stages. Anthropology, however, offers something even more important for the inductive verification of Darwinism than this relevant information.

Since the new theory is based on natural selection, it is quite evident that, in order to apply it to each separate case, a naturalist must have very detailed information on the various factors involved in the transformation of species. Such a requirement can be met only by the science of man. Indeed, while it took such an effort on Darwin's part to collect the most general and superficial facts on the birth rate and sex of domestic animals, comparable information with regard to humans is so plentiful that statistics record actual and relative birth rates of boys and girls. Instead of *habitata rara frequens, copiose, copiocissime,* and other terms by which the zoologist describes the concentration of animal populations in certain areas, the anthropologist operates with such detailed statistical data on population density that the zoologist cannot even dream about.

The same applies to the analysis of cases of migration, competition for survival, etc. It seems no accident that this theory is based strictly on the Malthusian doctrine, which was largely deduced from statistical concepts. Historical ethnography, on the other hand, offers us a very powerful tool in the analysis of these questions. It explains migration and mechanisms of replacing some populations by others, giving the sort of factual information that we have (although only in a much less complete form) for domestic animals and plants that accompany humans. Darwin refers to the supplanting of the black rat by the brown [throughout most of the world[3]] and of the black cockroach by the red [in Russia[4]] as examples of dramatic competition between related species. But how many more comparable facts are available in the human context!

In an earlier paper, I sought to stress the importance of human studies on the general theory of the origin of species. I did not, however, consider the other side of this question, partly because so much has already been said and written about it, but also because most people regard the application of the new theory to humans only as a debate on the origin of humans from animals rather than as a study on the very process of their origin, i.e. of the struggle, selection, and features fixed by selection. Prior to [the publications of] Darwin and Wallace, their followers Huxley and Vogt[5] attempted to apply Darwinism to the origin of man, but their aims were simply to show that humans, like other organisms, are related to simpler organisms. For this, they collected a number of zoological and anatomical facts linking humans to the higher apes in various respects. They merely repeated, with some additions, a review which had been carried out long before Darwinism, and done, in fact, much better.

Similarities between humans and apes has long been an important issue, discussed by many well-known naturalists, such as Tidemann, Vrolik, Grasiole and others. Thirty years before the publication of the first edition of Darwin's book,[6] Lesson was maintaining that with regard to their general constitution, apes are much closer to humans than to other animals; and it is only in their mental and judgmental abilities that they are far behind humans. Apes are indistinguishable from humans as far as their constitution and function of organs are concerned. He admitted, as did Tidemann, that "an orangutan brain differs from brains of other apes, but resembles a human one in many amazing details", etc. French academics, led by Isidor S.-Iler, also shared the opinion that the physical organization of humans and the higher apes was almost identical. Accordingly they justified their placement of humans in a separate systematic position based solely on psychological differences.

Thus, the followers of Darwin can lay claim only to a popularization of pre-Darwinian ideas, with some additions, including the proposal to return to the classification of Linneus (in which humans and apes are placed together as a section of the Primates) and the substitution of the old notion of anatomical affinity with that of consanguinity.[7] They merely proposed the term "general origin" instead of "general plan", without any consideration of how this general origin might have occurred. The founders of a new theory, however, went much further and presented their own opinions on the origin of man, a subject discussed in the next section.

II. WALLACE'S THEORY OF NATURAL SELECTION AS APPLIED TO HUMANS – SUMMARY OF DARWIN'S VIEWS ON ANTHROPOLOGY – SEXUAL SELECTION – PROBLEMS WITH THE THEORY OF NATURAL SELECTION – NATURAL SELECTION DOES NOT EXPLAIN THE ORIGIN OF MAJOR RACIAL DIFFERENCES

Wallace's papers on the descent of man caused a tremendous scandal among Darwinists. This very intelligent and courageous scientist did not confine his investigation to what had become a minor issue, namely the application of the new teaching [natural selection] to man. In his first paper (published in 1864), he did not clearly separate himself [from other Darwinists], but his view that natural selection in humans acts in an entirely different manner from that in the rest of the animal kingdom became evident in his second. Instead of changing the [physical] characteristics of organs to adapt them to the environment, natural selection influences the mental capacities of humans, inducing the invention of various means of adaptation to the external world. Animals which have exterminated an easy prey in one area must perforce acquire stronger legs and teeth, as well as greater strength and adroitness, by natural selection; otherwise they will not be able to hunt larger animals, and they themselves will suffer extinction.

Such a scenario is entirely inapplicable to humans: they do not need to acquire new physical abilities, which they would do extremely slowly, since instead of teeth and claws they can use bows and arrows, and they can keep warm with cloth and fire, rather than relying on hair and other organic contrivances. Thus mental development renders unnecessary an entire array of physical changes in the human body, preserving it in almost its initial shape. This, according to Wallace, explains the remarkable stability of the established human phenotype, as shown by a number of facts, including the observation that in the times of the pharaohs the human races were exactly the same as today.

Wallace's entire reasoning is based on the idea that progressive development of human mental capacity can occur incomparably faster and easier than corresponding changes in physical characteristics. If this were not so, there could be no explanation of human evolution following a different route from that of all other animals. This idea is expressed in a second, very famous article, entitled "Limits of Natural Selection as Applied to Man". Wallace here insists that savages have much bigger brains than is really necessary for their relatively simple way of life. Although the differences between the mental capacities of savages and modern men is dramatic, the former still have 5/6 the total volume of brain tissue of the most advanced races.

The main points of this conclusion apply not only to the lower races currently populating the earth but also to prehistoric man that lived at the time of the mammoth and cave bear. It follows that most of the brain tissue of these primitives was not essential for their way of life, but it was stored for the future to meet the needs of civilized man.[8]

This result just discussed cannot have been achieved by natural selection because it is not immediately applicable to the struggle for existence. Such development of the brain represents a luxury quite unnecessary for primitive man. Wallace supported these ideas with some analogous examples (including loss of body hair, in spite of the immediately harmful effects) and concluded that man did not originate as the result of natural selection but under the influence of a specific force. By moving from one far-reaching conclusion to another even more daring, he finally arrived at the ideas of Schopenhauer, proclaiming that any force results from a will, and that in fact "the entire world depends upon a will of superior reason, but represents this will by itself".

Darwinists, disappointed by the superficiality of Wallace's conclusions, did not pay sufficient attention to his major ideas concerning the relevancy of natural selection to humans and its limits, and what is particularly regretful, to his arguments on the inability of the selection theory to explain certain human characteristics. Darwin, for example, in mentioning the importance of the mental capacity of primitive man, notes in passing that he does not quite understand Wallace's contention that primitive man needed only slightly more brain tissue than an ape.

Instead of analyzing Wallace's theory on the relative disparity in mental development between primitive and civilized people, Darwin refers directly to a critical article by Klapared [2], to which we also now turn. In this article the entire question is transferred to the domain of deductive logic, arguing that man and other animals must obey the same general laws on the origin of species as other organisms. It follows, he says, if we postulate that superior reason took part in the origin of man, the same must apply to all other animals and plants. If, however, natural selection alone adequately explains the origin of the latter, there is no reason to make an exception for humans.

Turning the argument round, such statements by Klapared do not give adequate consideration to the foundations of Wallace's theory; in fact, he does not even analyze them. For example, he mentions Wallace's theoretical point that the abilities of primitive men are more than enough for their way of life but indispensable for those of civilized people, and adds: "I refrain from criticizing this opinion because the present [paper] is not about the defense of Darwinism". Could it be that this brief remark hides a simple inability to get the better of Wallace's ingenuous argument? Everyone would agree that the core of the heretical views of this famous naturalist-traveler [Wallace] is encapsulated in the opinion just cited, so this idea must be examined, regardless of whether it contains a potential vindication of Darwinism or not.

Klapared's assumed failure is illustrated by the reasoning in the following passage, which I consider sufficiently important to quote in full:

I, in my turn, would ask why Mr. Wallace does not apply the same reasoning to other cases. In the majority of birds, for example, the larynx is very complex and equipped with many muscles. It seems that all songbirds possess this particular characteristic, but many birds not notable for their singing also have a similarly well equipped larynx. This apparent contradiction can be explained on the assumption that such [non-singing] birds were previously able to sing, but, due to a variety of reasons, simply lost the singing habit.

If we attempt to apply this explanation to humans, we must say that the excessively large brain and unnecessarily large larynx of primitive and savage men, who cannot either sing or argue or count to more than five, are directly inherited from ancestors who could do all these things and many more, and they must have acquired a large brain and complex larynx solely through the slow process of natural selection. This assumption is so ridiculous that Klapared seemed afraid to make it, although it follows directly from his reasoning.

Lack of information prevents us giving a complete analysis of Wallace's arguments here. The basic features of the connection between mental development and the formation of the brain is still unknown, and much remains unclear about the structure and development of the vocal apparatus. Wallace's ideas, however, may be summarized in one general principle, which is easier to discuss than debating particular aspects. The assumption that primitives had larger brains than they actually required, as well as a more complicated vocal apparatus than was necessary for their speech and cries, can be reduced to the general idea of the recognition of *the capacity of the species* to make acquisitions beyond its needs. Such a principle is neither senseless nor meaningless. Facts that we observe daily in our economic existence are repeated in the economy of the rest of nature. We see generations of people accumulating wealth in quantities far exceeding their needs, whereas the initial impulse for this storage was to satisfy the most basic requirements. Enormous wealth flows from generation to generation and is hoarded in ever increasing quantities. When such inherited wealth comes into the possession of a particularly gifted, charitable, or otherwise remarkable person, who spends part of it on travel, research, charity, etc., it is finally used in a form encouraged by natural selection.

Let us offer another example, from a field more familiar to a naturalist. [Human] hair on the head and face is largely decorative and only partly functional, and it has therefore been acquired through the version of natural selection which Darwin calls sexual selection. The amount of hair grown by a man is, however, far in excess of that required by this form of selection. That is why hairdressing is one of the world's oldest crafts. Man sheared himself even when he was not able to dig and make fire, using the sharpened edges of shells and stones. A man with naturally long [uncut] hair would be regarded by everyone as eccentric or insane; even musicians and artists, for whom long hair is fashionable, cut it periodically. Thus, from the dawn of humanity, hair and nails have always grown excessively and wastefully, yet this excess could suddenly become very useful.

There is no need for further examples. Luxury exists in nature side by side with poverty, and extinction due to starvation coincides with the accumulation of huge surpluses. For example, nature produces too many males in many animals. Bees and spiders, as a result of a relatively well-developed brain in the former and a hunting instinct in the latter, mediate this luxury by exterminating males at the very time that they are no longer needed. Males of many animals, however, continue to live after that point in the absence of social and hunting instincts in the females. A farmer does the same as bees and spiders. He allows only the minimal number of males to live on his farm (since nature does not take care of this), and transforms monogamous wild animals (duck, canary, guinea-fowl) into domesticated polygamists. [3]

One might argue, nevertheless, that this interpretation of the facts is incorrect, because the principles of natural selection cannot be harmonized with the production of unnecessary, and even slightly harmful, excesses. I intend to give a detailed explanation of the polemics elsewhere, but I must point out that, contrary to this view, Darwinian theory is based on the acceptance of the existence of excesses in nature. Natural selection is a factor determined by the struggle for existence, i.e., *the struggle which originates from overabundant procreation.* If organisms did not procreate in "geometric progression" but in a less excessive manner, then the struggle, and hence the selection, would not be necessary. That is why Malthus advised moderation and abstinence to avoid the harmful effects of the straight struggle for existence. At the same time, excessive procreation, which generally provides the impulse to start the struggle, can sometimes be a powerful tool in that struggle. Many parasites, for instance, ensure their existence only by their enormous fertility.

Organisms in general compete for various surpluses. Animals are inclined to gluttony, the accumulation of excess food, etc. All this striving generates sources of struggle and thus clashes with natural selection.

It evidently follows that Wallace's arguments on reserves stored in the body of primitives are based on a reasonable idea: such reserves are indeed possible. His mistake is in interpreting the body's natural adaptability too narrowly.

Natural selection, says Wallace, could not produce an organ which is totally unnecessary for man, or the usefulness of which does not correspond to the degree of development of the organ itself. In other words, selection changes organs only to the extent that is absolutely necessary for victory in the struggle for survival, with no overdevelopment. Such an interpretation does not conform to the principles of natural selection established by Darwin and accepted by Wallace without reservation. Natural selection preserves and collects useful changes and eliminates harmful ones, but, precisely because of this, it has no effect on indifferent changes, which can be preserved. This point has been regarded as particularly important by botanists, and especially by Nägeli. They claim that a number of morphological features of plants are neutral, with no effect on survival (from the point of view of struggle and selection), but they are still extremely important for systematics.

This throws light on a hidden misunderstanding. Regarding natural selection as responsible for the elimination of neutral changes and having discovered such changes in the distinguishing characters of humans, Wallace comes, quite logically, to the conclusion that they may be produced by a force working in opposition to natural selection and much more powerful. The relevant changes in this particular case are neutral, because enhanced development of the brain and larynx, regarded as useless for primitives, is not harmful; it is, therefore, neutral with respect to the struggle for existence. Basically, this also applies to fat, which may exceed the real need for it but does not necessarily reach the harmful limits of obesity and degeneration, or excessive procreation, provided it does not follow a geometric progression and does not result in a struggle for existence.

Having presented a sketch of Wallace's views on man, we now turn to Darwin's anthropological theory, developed in much more detail, although, it seems, with less skill and talent.

In Chapter IV of *The Descent of Man,* Darwin discusses possible human development from various forms. By comparing humans to extant tetrapods, he attempts to show that natural selection can explain the enhanced development of the brain, mental capacities linked to adaptation of the anterior extremities [hands] for throwing and other special functions, as well as the adoption of an upright posture and walking. Almost any division of labor between organs improves performance and is therefore useful in the struggle for existence. Specialized hands, so dependent on the degree of intellectual development, not only require a change in the original posture so that walking is restricted to the legs, but the development of the head is also affected. When hands took over the function of attack and defense, the monstrous development of jaws, so typical of apes and other mammals, became unnecessary. For the same reason, canines became smaller and the teeth came closer together.

Although the major physical features of humans can be explained relatively easily as resulting from natural selection, the analysis should not stop here. Firstly, when the origin of the physical organization of humans is discussed, it must be constantly compared with its origin in the Quadrumana[9] [primates other than man], i.e. a comparison between families must take place. One of the major principles of Darwinism is precisely that selection works much more efficiently in smaller systematic groups, and that is why it is so important to compare individuals of the same species or similar species of the same family. For the same reason, a limited comparison should be made between species which show a much greater degree of similarity than that between man and other primates, with the aim of achieving a complete and correct placing of the origin of man under the laws of natural selection. Such a goal is, however, impracticable because there is no known intermediate between humans and other higher primates.

The second reason why such a profound and meticulous scientist as Darwin does not provide such an explanation of human physical features is that it takes into account only the most obvious characteristics. For example, absence of hair is a feature that cannot be ascribed to the effects of natural selection, and Darwin invents sexual selection to include it.

Darwin summarizes his view on the origin of man from the lower forms at the Conclusion to Chapter IV in the following statement:

As all animals tend to multiply beyond their means of subsistence, so it must have been with the progenitors of man; and this will inevitably have led to a struggle for existence and for natural selection. This latter process will have been greatly aided by the inherited effects of the increased use of parts; these two processes incessantly reacting on each other. It appears, also, as we shall hereafter see, that various unimportant characters have been acquired by man through sexual selection. *An unexplained residuum of change, perhaps a large one, must be left to the assumed uniform action of those unknown agencies, which occasionally induce strongly-marked and abrupt deviations of structure in our domestic productions* [italics added by Metchnikoff]. [4]

Darwin, like Wallace, reached the limits of natural selection, but while Wallace does not cross the borders of physical appearance, Darwin does everything possible to include humans within a broader version of his theory of the origin of species. He inevitably has to turn from general studies of physical features to an analysis of human races, their meaning and origin. While Wallace only touches upon this question, which is very significant to the theory of Darwinism and most important for anthropology, Darwin devotes more than half of his *The Descent of Man* to it. In Chapter VII he discusses, with remarkable objectivity, the question of the influence of different factors on racial features, and comes to the conclusion that

the characteristic differences between the races of man cannot be accounted for in a satisfactory manner by the direct action of the conditions of life, nor by the effects of the continued use of parts, nor through the principle of correlation. [5]

The same negative conclusion is reached with respect to natural selection, since such pressures produce only those features which are immediately useful for the organism, while

not one of the external differences between the races of man is of any direct or special service to him. [6]

He concludes:

We have thus far been baffled in all our attempts to account for the differences between the races of man; but there remains one important agency, namely Sexual Selection, which appears to have acted as powerfully on man, as on many other animals. I do not intend to assert that external selection will account for all the differences between the races. An unexplained residuum is left, about which we can in our ignorance only say, that as individuals are continually born with, for instance, heads a little rounder or narrower, and with noses a little longer or shorter, such slight differences might become fixed and uniform, if the unknown agencies which induced them were to act in a more constant manner, aided by long-continued intercrossing. Such modifications come under the provisional class... [7]

To clarify the extent and mode of action of sexual selection in humans, Darwin first lays down some general principles for this precept, then gives an overview of the so-called secondary sexual characteristics of animals, and finally turns to the application of the available results on racial differences. This theory is of such crucial importance to anthropology that I shall devote much of this text to it, particularly because it has never before been subjected to a detailed analysis.

Elaborating his main idea that man is descended from ape-like ancestors, Darwin presupposes that the mating customs of primitive man were similar to those observed among living non-human primates. It is known that some apes live in pairs throughout the year, others form pairs only for a restricted part of the year, and still others, like the gorilla, are polygamous. In all these cases we see one or another form of mating, but not a random mixing of the sexes that many authors ascribe to humans in the early stages of their evolution, but which Darwin considers to be a relatively late development.

In his opinion, the behavior of primitives was governed by instinct, rather than by reason, and that is why they did not mate too early, did not kill their children, and, in general, lived like higher animals rather than present-day savages. Therefore sexual selection must have been much more powerful in ancient times than now. External characteristics would have been the most important criteria [for selection of mates], rather than wealth and social status, which are so important for marriage among civilized people. Strength of males and beauty of females, therefore, must have been the features most subjected to sexual selection in the primitive era, but while strength is more or less the same for all people, what constitutes beauty varies widely. Today, it is well known that people consider the external features of those of their own race or country to be the most beautiful. For example, a wide round face, narrow eyes, prominent cheek bones, and a flat nose epitomize a model of beauty for Mongols, just as [they regard] a flat steppe as a beautiful landscape, etc.

Let us suppose the members of a tribe, in which some form of marriage was practiced, to spread over an unoccupied continent; they would soon split up into distinct hordes, which would be separated from each other by various barriers, and still more effectually by the incessant wars between all barbarous nations. The hordes would thus be exposed to slightly different conditions and habits of life, and would sooner or later come to differ in some small degree. As soon as this occurred, each isolated tribe would form for itself a slightly different standard of beauty; and then unconscious selection would come into action through the more powerful and leading savages preferring certain women to others. Thus the differences between the tribes, at first very slight, would gradually and inevitably be increased to a greater and greater degree. [8]

The effects of sexual selection in humans must be significantly strengthened by the fact that it is not a one-sided but a two-sided choice: not only does a man choose a female, but *vice versa*. This results in a combination of the strongest and most attractive males with the most beautiful females, and consequently the offspring of such marriages live longer than their weaker fellow-tribesmen.

Due to this two-way selection and the fact that changes are inherited from generation to generation, it is difficult to determine which [sex] first acquired any particular feature. For example, Darwin considers it quite likely that loss of body hair was first [selected] as a form of female decoration, and thereafter was inherited by both sexes [9]; but whereas the female lost her body hair and beard, the male kept the latter and even transformed it into an important sexual feature developed through sexual selection. Darwin also suggests that a beautiful and gentle voice is a characteristic that initially appeared in females.

It should not be overlooked (as P. Broka did in his review of the theory of sexual selection) that Darwin includes humans within [the scope of] this process, on the one hand because sexual selection undoubtedly induces the [first] appearance of many distinctive features in various animals, and, on the other hand, because of the difficulty of explaining racial differences in any other way. If we admit, says Darwin, that sexual selection removed the hair from an ape's buttocks and painted the cheeks of a male mandrill, then why not suppose that the same factor bared the entire human body and painted it this or that color.

The end of Chapter XX gives the best possible illustration of Darwin's reasoning on the applicability of sexual selection to humans.

The views here advanced, on the part which sexual selection has played in the history of man, want scientific precision. He who does not admit this agency in the case of the lower animals, will properly disregard all that I have written in the latter chapters on man. We cannot positively say that this character, but not that, has been thus modified; it has, however, been shown that the races of man differ from each other and from their nearest allies among the lower animals, in certain characters which are of no service to them in their ordinary habits of life, and which it is extremely probable would have been modified through sexual selection. We have seen that with the lowest savages the people of each tribe admire their characteristic qualities – the shape of the head and face, the squareness of the check-bones, the prominence or depression of the nose, the color of the skin, the length of the hair on the head, the absence of hair on the face and body, or the presence of a great beard, etc. Hence these and other such points could hardly fail to have been slowly and gradually exaggerated from the more powerful and able men in each tribe, who would succeed in rearing the largest number of off-spring, having selected during many generations as their wives the most strongly-characterized and therefore the most attractive women. For my own part I conclude that of all causes which have led to the differences in external appearance between the races of man, and to a certain extent between man and the lower animals, sexual selection has been by far the most efficient. [10]

Darwin tells us so directly how he views the application of the theory of natural selection to humans, and how it should be viewed, that I am surprised and indignant at critics like Bastian [11], who can only crack vulgar jokes at the great thinker's endeavors to shed light into an area which, until now, has been completely pervaded by outer darkness. Bastian, whose works fill many scholarly volumes, offers an unclear and not very scientific argument based on the claim that Darwin often departs from the facts and is carried away into a field of potentialities. A narrow-minded critic does not understand the importance of hypotheses that do not follow directly from facts but which are still consistent with them and explain them, and that is why Bastian does not consider it necessary to peer into the foundations of the sexual selection theory. Instead, he reiterates a modernized version of an old and long-refuted theory of the origin of races as a response to the immediate influence of environmental factors. This theory, which regards racial differences as arising from the effect of climate, food, and other external agents, was subjected to scientific scrutiny in the last century by Pallas, whose conclusion regarding it was entirely negative. Subsequently, many scientists have sought to appropriate this theory, particularly under the patronage of the so-called monogenists. Several detailed analyses of this theory have been instituted, and each time it has been virtually destroyed by its critics (e.g. by Gordon in his famous book *On Species and Races*). Even such a monogenist as Veintz, who gave a conscientious analysis of this theory, came to the conclusion [12] that, although the color of skin is partly determined by climate and way of life, it also depends on heredity, and this is also applicable to hair-color and hair-type.

It has long been known that the entire [North] American continent was populated by just one race, in spite of the vast range in climate, while tropical Africa and tropical America were populated by different races [from the natives of North America]. It is also known that the most densely colored Negroes, the Iolofes, are bordered to both the north and south by less densely colored peoples, namely Hametic tribes and less pigmented Negroes. In spite of these scientific facts and many more (Schveinfruit recently described two Central African tribes, the Niam-Niam and the Aka, whose skin is much paler than that of genuine Negroes), Bastian [13] dares to repeat that skin color depends directly on climate, arguing that "the Portuguese in Gambia have been turned into real Negroes"(!) [*Die Portugiesen am Gambia sind (nach Demaret) zu Negeru geworden.*] Hair type, so distinctive in many races, is written off by him as due to electricity in the air, etc. It is evident that "facts" related by travelers are accepted uncritically by Bastian and are regarded as superior to any hypotheses, even the most brilliant. In Madeira and, later, in Lisbon, I used to meet adult Portuguese and children who lived on the west coast of Africa and the Cape Verde Islands, and who bore not the slightest resemblance to Negroes. I do not, of course, refer to mulattos, whom I happened to see quite often, since they result from repeated crosses between Portuguese and Negroes.

Putting aside critics like Bastian, I consider it necessary to turn to an analysis of the theory of sexual selection as applied to humans based on major principles proposed by Darwin himself.

I. For sexual selection to be able to initiate races, it is important that different groups of the population have different ideals of beauty, and that sexual selection takes into account such complications as aberrant sexual behavior ([e.g.] communal marriage), infanticide, and early betrothals. We may find examples of these practices among living savages or semi-primitives. However, in spite of the existence of certain forms of marriage and the absence of early betrothals and infanticide, there is such promiscuity that one [ideal of beauty] is frequently replaced by others.

Let us reinforce this by examples. Even though Negro and Caucasian races represent utterly different [racial] types, various mixtures between them have constantly occurred and still occur today. Everyone knows that very many mulattos live in both Americas, and in Africa, where European people come into contact with the Negro race. The latter, however, mixes far more with less civilized Caucasians. Wandering in East Africa, "a conscientious traveler", says Mutzinger, "is never able to determine where the real Negroes live, and his belief in any strict differentiation of races becomes more and more elusive." Transitions between races are considered to result from such mixtures [14], and there is even a recent example of this. According to Abyssinian legend, the Gallas are descended from an Abyssinian woman and a black slave; inconstancy in physical features in this tribe clearly reflects such a mixed origin.

A neighboring population, of more recent origin, is thought to be mixture of Gallas and Arabs. According to their own legends, their patriarch Isaac moved from southern Arabia and married a Galla woman. This legend can be explained only by a recent migration of Arabs to Somalia, [which was] occupied by Gallas, to produce new Somalis, the Baung [15]. It is known that they have a definite form of marriage, that the women get married as early as 13 [16], that wives are purchased by dowry, not promised in childhood, and, finally, that the practice of infanticide has not spread there.

Throughout West Africa, the aboriginal and Arab racial types have intermixed to varying degrees. Racial mixtures have also occurred in East Africa: nationalities known as Torodos, Jao Jalonki, Ssissilbe, Jasambe, and Soromasa represent mixtures between Negroes and Fulgas, who invaded the lands populated by Negroes as conquerors and proponents of Islam [17]. It should be mentioned that neighboring peoples, living side by side, quite obviously despise each other. This is well known with regard to Europeans and Negroes in America, and applies to the same extent to both Gallas and Fulgas, who disdain Negroes. One can conclude from this that the motives which drive people to interbreed are more powerful than mutual hatred and human ideals. It is therefore not surprising that peoples of mixed origin are so numerous. These include Melanesians and many Polynesians, of mixed Malayan and Papuan descent; Magyars, a mix of Finnish, Slav and German tribes; Russians; present-day Turks, who combine Greek, Armenian, Slav and Circassian blood, etc. It should also be mentioned that these examples relate to mixtures of races and peoples with established [racial] types, while sexual selection acts to separate races when the characteristics of these races make their first appearance.

The general conclusion from the above (as far as we can judge from these recent cases) is that the differentiating force of sexual selection can not overcome those instincts that lead to the mixing of races. [18]

II. According to the scheme proposed by Darwin for understanding the action of sexual selection in humans, [changes in] females have been largely towards the development of elegant features, while those in males have been primarily directed towards power, fortitude, energy, and the like. In the Summary to Chapter XX he says:

We may conclude that the greater size, strength, courage, pugnacity, and even energy of man in comparison with the same qualities in woman, were acquired during primeval times, and have been subsequently augmented, chiefly through the contests of rival males for the possession of the females. [19] It is not improbable that the females were modified in other respects for the same purpose and through the same means; so that women have acquired sweeter voices and become more beautiful than men. [20]

Since, according to Darwin, racial features are nothing else than a fixed embodiment of different ideals of beauty, and since an acquisition of the latter fell to women's lot, then it becomes evident that it was precisely women who revealed racial features first and who handed them down to their children of both sexes. Darwin never expresses this opinion openly, but it clearly follows from his previous discussion in Chapter XX:

"Other characters proper to the males of the lower animals, such as bright colors and various ornaments, have been acquired by the more attractive males having been preferred by the females. There are, however, exceptional cases in which the males, instead of having been the selected, have been the selectors. We recognize such cases by the females having been rendered more highly ornamented than the males – their ornamental character having been transmitted exclusively or chiefly to their female offspring. [21] Women, however, certainly transmit most of their characters, including beauty, to their offsprings of both sexes; so that the continued preference by the man of each race of the more attractive women, according to their standard of taste, would tend to modify in the same manner all the individuals of both sexes belonging to the race." [22]

An exception to this is represented by a beard as a racial characteristic acquired by males through sexual selection.

An inescapable deduction from this conclusion is determined by the fact that, in a majority of cases regarding humans, a selection is carried out by males, i.e., contrary to that which Darwin considers to be a general rule for animals.

Man is more powerful in body and mind than woman, and in savage state he keeps her in a far more abject state of bondage than does the male of any other animal; therefore it is not surprising that he should have gained the power of selection. Women are everywhere conscious of the value of their beauty; and when they have the means, they take more delight in decorating themselves with all sorts of ornaments that do men. They borrow the plumes of male birds, with which nature decked this sex in order to charm the females. [23]

Notwithstanding the general importance of females in the production of racial differences, man, like the males of many animals, shows much higher variability than woman. This reality, expressed by Darwin himself [24], is one of the rare positive conclusions of anthropology, and accents the case that, in the formation of racial or other secondary differences, males play quite a significant role. If this were not so, humans would form an exception among animals, not only because of their unique method of choice (by males, not females) but also because of the variability of their features. It is known that females show more stability [than males], or, as Rutimeyer puts it: "Representatives of the female sex, in both their individual development and genealogical history, are less likely to deviate from the primary type than those of the male sex." If we compare this rule, the correctness of which has also been recognized by Darwin, to the latter's views on the distinctive racial differences between men, then we see that the application of the theory of sexual selection to the origin of races faces an insoluble contradiction.

III. Darwin, being unable to explain the origin of racial differences to recent factors, transposes the entire process to an earlier period. He draws attention to a number of elements that immobilize sexual activity in contemporary humans, like early betrothals, infanticide, etc. The same result may be achieved by civilized people through marriages based on motives other than external beauty, like material advantage, etc.

Thus during these primordial times all the conditions for sexual selection would have been much more favorable than at a later period, when man had advanced in his intellectual powers, but had retrograded in his instincts. Therefore, whatever influence sexual selection may have had in producing the differences between the races of man, and between man and the higher Quadrumana, this influence would have been much more powerful at a very remote period than at the present day. [25]

Here again a significant contradiction occurs. If sexual selection itself becomes powerless when savage man develops to the state of contemporary man, then its results must be revealed in the production not only of racial differences but of secondary sexual features also. Meanwhile, the opinion that "secondary sexual features are shown more vividly in peoples at a higher stage of development" [26] than in those less civilized, has been quite firmly established in science, and is based on numerous and mostly accurate data. On a smaller scale, some regard comparisons between [people at] higher and lower cultural levels within the same community as illustrating the same point. Thus, for example, Retzius found that, in Sweden, sexual differences between the skulls of people in higher social groups are much more clearly marked than in peasants. Proceeding from the idea that a model of beauty can be acquired only when Reason [mind] is sufficiently developed, P. Broka [27] comes to the conclusion that "sexual selection is unlikely to influence the life of savages, but it acquires significance, and it grows, together with the progress of society."

It clearly follows from these statements that, not only is sexual selection not suppressed by the progress of civilization, but it grows with it, and it becomes more and more influential, affecting not only the divergence of secondary sexual characteristics but also, to some extent, the formation of racial features. This, however, does not occur in reality.

IV. Thus, even though we are convinced that the actual principles of sexual selection are correct, we must not try to use this factor to explain racial differences because of all the contradictions just mentioned. Any attempt to eliminate these contradictions, if at all possible, should deal with another obstacle to explanations in terms of sexual selection. This is that many racial features are too deeply rooted in the organism to be affected by sexual selection, even if the latter is influenced by environmental factors.

Because the majority of racial characteristics are stamped on the external features of man, it is easy to come to the conclusion that voluntary choice is the prime initial factor for the acquisition of these factors. A number of facts reflect the considerable attention paid by savages to, apparently, the most insignificant features of their external appearance. Who among present-day [civilized] humans bothers about the shape of his head and whether some [people] have longer [heads]? Savages, however, consider an oval head to be the ideal shape and feel it necessary to establish this shape artificially in their children.

Schweinfurt describes the following situation from his travels in Central Africa. Once, a misunderstanding arose as to which tribe, Divka or Bongo, a certain boy belonged. To resolve this dispute, one of Schweinfurt's interpreters examined the boy's head, and established that he was certainly born of Bongo parents because his head was broad. "Our women," explained the interpreter, "follow the custom of squeezing the head of a new-born baby from top to bottom, while Divka women squeeze it from side to side." Everyone knows to what extent primitives are conscious of their coiffure, tattoos and other devices to acquire perfect beauty. According to Bakker, natives of North Africa take up to 10 years to finally complete their coiffure, and the complete tattooing of Mauris is not only incredibly lengthy and laborious but also so dangerous that death sometimes ensues.

Skin color, facial features, type of hair, etc., illustrate racial characteristics that are easy to recognize and which, for this very reason, are subject to human choice and cultivation. Precisely because of this, it is easy for Darwin to conclude that they have been developed through sexual selection. Indeed, if humans have been changing breeds of domestic animals and plants for so long, just to satisfy their long-established aesthetic taste, then why not suppose that they might subject themselves to the same process, and for the same purpose? But, as mentioned above, racial features are rooted much deeper in the macrocosm of the human body. Not only the [overall] form of the human body, but also the proportions of different parts of the body are important in the consideration of racial features.

The lengths of the extremities and of the head in proportion to the body can be very important diagnostic features for certain races, and this is noticeable even with smaller organs. For example, [it is claimed that] the arm [above] the elbow is much longer in relation to the forearm in Negroes than in Europeans, a point that has often been emphasized by researchers intent in pointing out similarities between Negroes and apes. To establish such ratios, anthropologists must make a number of precise measurements, and these show that, in spite of their theoretical importance, such ratios provide less obvious differences than factors such as skin color or facial features. If it is rather difficult to imagine that the relative sizes of bodily parts can be the subject of sexual or any other form of artificial selection, then it is even more difficult to concede this with respect to less obvious racial characteristics. It is known, for example, that skulls differ not only in length, width, and height, which are easily observed in a superficial examination, but also in various features of the base of the skull, which can be examined only in a prepared skull. Recently Ekker called attention to the fact that the base of a Negro's skull is longer than that of Europeans, and therefore the hole at the back [(the foramen magnum)] is positioned differently. At the same time, he pointed out that this feature places Negroes closer to animals, and it represents a factor of primary significance in identifying races from skulls.

Another fact to which I should like to refer concerns differences in the structure of the pelvis. For a long time many authors tried to prove that this part of the skeleton shows marked dissimilarities in different races, and insisted, for instance, that the Negro's pelvis shows resemblance to that of an animal. More careful observers regard these as hasty, untenable conclusions, but even they admit that the shape of the pelvis varies between different races, in some cases quite significantly. One of the most recent scholars in this field, F. Frank, whose critical approach to widespread opinions on racial differences of the pelvis is well known, still implies that there are such pelvic differences between Malaysians and Negroes.

Burmeister, the famous German naturalist, lived for a long time in Argentina and took the opportunity to study Negroes closely. He claims that not only the skull and the pelvis but also other parts of the Negro's skeleton, down to the smallest bones, show varying degrees of racial difference.

Another set of facts that emphasize the independence of racial features from [sexual] selection can be expressed in the general statement that racial features change with age, and that the most pronounced changes do not coincide with the onset of sexual maturity. Quite naturally, if sexual selection facilitates the development and establishment of racial features, then the latter must be at their peak of development when humans achieve marital age. It is known, as mentioned by Darwin, that the distinguishing features of the males of different animals, like the beautiful plumage of birds, make their first appearance close to the beginning of sexual maturity. Such coincidences are the only means of confirming the occurrence of sexual selection. It should therefore be expected, from a Darwinian point of view, that human racial features should reach their peak development at about the same period [of sexual maturity], and the same should be true of secondary sexual features.

The facts, however, do not support such assumptions. The physical features and voice of human males develop at the onset of sexual maturity, but the beard and mustache grow several years later. Racial features, nevertheless, are revealed much earlier: the black skin of Negroes develops within the first year in the Sudan, and by three years in Egypt, and curly hair appears even earlier. According to Darwin, some features result from direct competition between males, while racial differences are acquired by females through selection and then inherited by both sexes. Those who seek to explain the lack of synchrony of secondary sexual features which are obviously dependent, one factor, sexual selection, must also be considered relative to other facts. For example, some of the characteristic features of Mongols appear in early childhood, diminish to some extent in maturity, and almost completely disappear in old age. Such [examples include] facial peculiarities, i.e. the most obvious features. I refer firstly to the characteristic shape of the eye [28] and the somewhat flattened nose. To a lesser extent, but within the same category, are the beard and mustache of Mongols, and also those of Negroes. These are completely absent in adolescence, the time when most marriages occur in both Mongols and Negroes. The beard and mustache grow significantly later, when sexual selection is of less significance.

Summing up all the aforesaid, it is easy to reach the conclusion that sexual selection, in the form established by Darwin, does not form a basis for the explanation of the major problems of anthropology. It leads to a number of more or less intricate contradictions, and does not hold the key to the understanding of racial differences. It is certainly not my intention to say that sexual selection has not played any role in the anthropological development of humans, but I do suggest that its influence is quite limited, while, on the contrary, the domain of "unexplained facts", to which Darwin refers at the end of Ch. VII, is very extensive.

III. THE STRUGGLE FOR EXISTENCE AND SELECTION IN HUMANS – THE MANNER OF NATURAL SELECTION. ITS RELATION TO MORPHOLOGICAL CHANGES – THE ROLE OF NATURAL SELECTION AS A CREATOR OF SPECIES MUST BE LIMITED

In the previous section I analyzed, as thoroughly as possible, Darwin's attempts to explain major anthropological problems from the point of view of the new theory, and the decisive move of Wallace, designed to escape from these complications. It clearly follows from what has been said that, in spite of the efforts of the two greatest naturalists of our century, the results are quite modest. Neither natural selection in the general sense, nor the particular form [known as] sexual selection, are capable of resolving the major problems of the origin of races. They merely shed some light on the problem of the descent of man. And here, quite naturally, the reason for such a failure arises.

Let us remind ourselves of the major principles of Darwinism from which the evolution of species may be logically deduced: 1) all organisms produce more offspring than can survive; 2) this generates a struggle for existence, in which the strongest, i.e. those most adapted to the environment, survive; 3) in this way, a series of seemingly insignificant individual differences are selected, and organisms without such advantages become extinct; 4) these [differences] thus become widespread and are preserved by heredity to create new species or varieties.

Let us now consider how these principles apply to humans. Darwin borrowed [the idea of] the over-reproduction of organisms leading to competition from Malthus, who developed his theory precisely with regard to humans. Darwin himself says several times that his theory is "the doctrine of Malthus applied to the whole animal and plant kingdoms", and even frequently uses the graphic, but incorrect, assertion that [unrestrained] procreation results in "geometric progression". "The struggle for existence" in humans occurs in two very different forms: between individuals of the same tribe or race, and between representatives of different races. There is no need for further discussion on this subject, but, in the present context, it would certainly be useful to take a closer look at the struggle between races and its principal results.

The following extract is from the analytical part of F. Müller's textbook [29]: "Taking into account all the data, one can conclude that the three most numerous races, Mongol, Mediterranean, and African Negro, have the best chance to win the struggle for existence, which has recently intensified more than ever". Probably no one would attempt to draw a connection between success in this struggle and those racial differences that are widely known and which serve as distinctive characteristics of the races, such as the black skin and curly hair of Negroes, and the flat face and straight hair of Mongols, etc. Nevertheless, I must still demonstrate that natural selection is not associated with these morphological characteristics of the races.

Among every race there are groups of people that survive and [others] that become extinct. Among Mongols, for instance, some strictly Mongolian peoples have become extinct, such as the Samoyeds, Ostyaks, and other small tribes, which happen to be the most characteristic representatives of this race. The Chinese, Japanese, Yakuts, and others have not only survived but [their survival] is somehow related to the extinction of other groups. It becomes clear, even from this example alone, that the racial factor is of much smaller significance for possible extinction than the cultural one, like, for example, the degree of "civilization". We can still distinguish some quite numerous and powerful tribes among native Americans. This leads to the conclusion that [certain] Indian tribes in Central and South America not only resisted extinction by Europeans but even forced the latter out, revealing considerable energy and mental capacity. The vitality and strength of certain Malayan peoples (e.g. the Yavans) whose population has significantly increased since they came into contact with Europeans, is well known. On the other hand, many Caucasian tribes have vanished, and still others are in the process of extinction. The Guanchs, a Hamitic race which many researchers consider to be related to the Basques, have quite recently disappeared. These latter people, widespread in Europe until recently, remain only in the Pyrenees, but they risk extinction there as well, because of the endless Carlist rebellion.

There is not the slightest doubt that purely ethnic features, i.e. distinctive character, different customs (infanticide, cannibalism, blood feuds, etc.), degree of civilization, as well as certain physiological factors, such as higher or lower fertility, play a major part in people's propensity to survive or become extinct. Within one and the same region, however, the most diverse ethnic groups can live side by side. We see Mongols next to Chinese, Circassians with Jews, Arabs with Berbers, etc. It is clear, therefore, that natural selection, dividing humankind into winners and losers in the struggle for existence, is governed not by some obvious differences in form but by far deeper ethnic, and partly physiological, features.

So, excessive procreation, struggle for existence, and selection (which leads to the survival of the fittest and to the extinction of the less well adapted) takes place in all humans as a species. As to richness in individual peculiarities, humans are not only the equals of other animals but are even ahead of them. Therefore all the prerequisites for Darwinian evolution (by means of natural selection) seem to be in place, but nothing of this sort has yet been noticed [30]. On the other hand, we see a number of varieties or races within the species *Homo sapiens*, but we cannot explain their origin in terms of either natural selection or sexual selection.

In order to appreciate the significance of this incongruity, we must turn to other organisms. In developing his theory of the struggle for existence, Darwin establishes that this struggle is most intense between organisms that are closest in structure and function, i.e., between individuals and varieties of the same species, or between similar species of one family. He supports this presupposition with the following examples. One type of swallow supplanted another from the whole of the United States of America; one type of thrush (*Turdus viscivorus*[10]), due to more successful breeding, significantly reduced the population of another type of thrush (*Turdus musicus*[11]) in certain parts of Scotland; the red rat supplanted the black; one type of cockroach expelled its bigger relative in Russia; the European [hive] bee, exported to Australia, notoriously exterminated a smaller, stingless, native bee; and, finally, there is the case of one type of mustard supplanting another.

"We can dimly see," adds Darwin, "why the competition should be most severe between allied forms, which fill nearly the same place in the economy of nature; but probably in no one case could we precisely say why one species has been victorious over another in the great battle of life."[12]

It is very important that we now establish whether the relative success of each of the above-named species is dependent on their diagnostic features or not. If, as the theory argues, distinctions between species arose because of their usefulness in the struggle for existence, then it is natural to conclude that the conquering and supplanting of one species by another results from the occurrence of particularly useful characteristics.

So, what are the differences between the species-winners, enumerated by Darwin, and their defeated relatives? The European bee differs from the Australian (which belongs, in fact, to another genus, *Melipona*) by its greater size and the presence of a sting, whose significance in the struggle for existence is beyond doubt. Other cases, however, are quite different. Thus, the mistle thrush, the winner, has such insignificant distinctions from the song thrush, the loser, that they are hardly noticeable at first sight.[13] Both have almost identical coloration, but the feathers covering the ventral abdomen of the first are blackish-grey, while they are rust-colored in the second; two transverse stripes on the wings of the first thrush are rather lighter than those of the second, etc. It is well known that the coloration of animals, especially birds, often present features of the utmost significance in the struggle for existence: green is very useful for animals living on the green parts of plants; gray is important for those that frequent the dry parts of trees or fallen leaves. While it is easy to agree with this opinion, it is impossible to accept that slight differences in coloration, like those that distinguish the two types of thrush, can play a role in the struggle for existence. The same applies to cockroaches. Longer wings, lighter coloration, two dark stripes on the chest, and other small distinctions of the same sort cannot explain why the red cockroach has forced out the black. In the same way, the straight hair, copper-red skin color, flattened nose, and other features of the Malayan cannot explain his advantage in the struggle for existence over the curly-haired, black-skinned, and acquiline-nosed Papuan.

There is no clearer case in the whole of zoology of one species supplanting another than that of rats. The brown rat, the so-called pasyuk, moved to Europe during the early third of the last century; it appeared in America in 1775. Despite its recent arrival, it has spread almost throughout the world, expelling its predecessor, the black rat, which had populated Europe at least since the Middle Ages. Here are the distinctive features of each species:

"Pasyuk [= Brown Rat], *Mus decumaas*.[14] Body long; ear not reaching eye when pressed to head; reddish-gray, white ventrally; length 7.5 inches, tail 7 inches.
[Black] Rat , *Mus rattus*.[15] Ear covering eye when pressed to head; tail longer than body; reddish-black with greenish tint dorsally, lighter grayish-black ventrally; length 6 inches, tail 7.5 inches."

Of all these characters, only size is important, although the least significant from a zoological point of view. Shorter ears do not cause any inconvenience to pasyuks, nor, it seems, hinder them in expelling black rats. Features, such as the slightly shorter tail (with 210 rings, as opposed to 250–260 in the black rat) or the 47 roughnesses of the skin folds, give the pasyuk no obvious advantages over its rival, but its ability to withstand larger amounts of water and its greater boldness and hardihood must play a more significant role.

These examples reveal the essence of the whole issue, i.e. the conclusion that *the most important factors in the struggle for existence are not the insignificant morphological distinctions between species but their physiological characteristics, such as greater fertility, strength, endurance*, etc. Morphological and physiological distinctions frequently occur together, but this is not a regular coincidence. Closely related species, from a zoological point of view, may well have different strengths in the struggle for existence. Although little is known about the breeding productivity in different animals, we can already say that species classified as closely related may invest different amounts of material in the production of new generations. Thus, for example, the eggs of a hawk represent 13% of its body weight, while the figure for the buzzard (*Buteo vulgaris*) is 23.5%. A corresponding example for closely related fish is provided by *Cyprinus jees* with [eggs representing] 26% and *C. baleris* 36% [31].

Adopting this position allows us to explain some facts that puzzled Darwin. Putting forward the concept of the [relative] superiority of [certain] forms, he mentions, for example, that animals and plants exported from Britain to New Zealand supplanted many native species, but no one noticed the reverse process, i.e. the spreading of New Zealand species in Europe at the expense of native European forms. From this point of view, one can say that British products are at a much higher level on the ladder of living organisms than their New Zealand counterparts. This result, however, could not have been predicted by even the most experienced naturalist, even after the most careful investigation of the species of both countries. In other words, strength and fitness in the struggle for existence do not coincide with the morphological features by which a naturalist distinguishes and classifies the same species.

This rule, deduced from the study of differences between human races and [between] related animals and which provide examples of selection in the struggle for existence, has a wide range of applications to the classification of organisms. However, difficulties arise in discussing any one animal or plant in determining the benefit or harm resulting from any particular feature. We can, nevertheless, make some quite probable predictions if we remain within the anthropological data. Let us consider some examples. The group of sponges is made up of many forms of simple organization, which live almost entirely on the sea bottom. These animals appeared as early as the Paleozoic epoch and are still quite widespread in [modern] seas and thus represent one of the most fundamental groups. Recent careful studies show that not only species but also families can be identified through careful microscopic analysis of the smallest spicules that make up their skeletons. For example, the *Suberites* family is characterized by the presence of pin-shaped spicules only, while in *Esperia* these appear together with peculiar hooked spicules. *Reniera* has spindle-shaped spicules. The differences distinguishing species seem even less significant. *Asceta primordialis*, for instance, has three-pointed spicules with sharp tips, while in *Ascandra* they are relatively blunter; rod-shaped spicules in *Ascandra pinus* are 4–6 times as long as the three-pointed ones, and the main axis of the latter is longer than the laterals; rod-shaped spicules in *Asc. variabilis* are only 2–3 times as long as the three-pointed ones, and their main axis is shorter than the laterals, etc.

What is the significance of such microscopic distinctions in the life and struggle for existence of larger animals, whose external form is so variable that it may not provide a basis for classification? Here, as in the case of human racial features or specific distinctions between red and black cockroaches, we are apparently dealing with neutral characteristics, i.e. those that are not immediately useful in the struggle for existence, but are very important morphologically; they are independent of natural selection, but can be used in the classification of species, as in sponges. Let us consider another example, also taken from the realm of lower animals. There are several species in the group of infusoria, the Paramecina. Among them, a species of *Paramecium* is distinguished by the very deep mouth rim, the mouth elliptical and twisted through 50°, and a short hair in the gullet; a related species of *Colpoda* is distinguished by the mouth situated in a marked depression, with a cluster of long cilia on its lower edge. Quite possibly these longer cilia transport food to the mouth better than the short ones found in *Paramecium*. But how insignificant (in relation to the struggle for existence) this distinction seems when compared to the capacity of *Colpoda*, but not *Paramecium*, to cocoon itself when drought or other unfavorable conditions occur! This feature leads to the widest possible spread of *Colpoda*, while *Paramecium*, which cannot withstand drought, can live only in permanent puddles of water.

I mention here only lower animals in order to reinforce the conclusion that there is no correspondence between the morphological peculiarities and the physiological advantages of various species (and, to some extent, families) to win the struggle for existence. This idea could be developed further by analyzing, one by one, each simple organism, listing even the slightest differences between them relevant to their coexistence (like differences between shells, spicules, and other skeletal formations) as well as the possession of such powerful weapons in this struggle as enormous fertility and amazing vitality and endurance. It is these latter characteristics that have enabled such animals to spread throughout the earth and survive from the most ancient times to the present.

Nägeli, one of the most remarkable critics of Darwinism and a proponent of the theory of succession of species, noted long ago that

> useful adaptations cited by Darwin with respect to animals, and quite possibly common in plants, represent qualities of a purely physiological nature and indicate the development or transformation of that particular organ for a certain specific function. I am not aware of the existence of even one morphological development that can be explained by means of the theory of usefulness (i.e. natural selection), and I cannot understand how such changes might occur, since morphological features in general are quite indifferent to physiological function. [32]

What Nagely has in mind here is the specific positioning of cells, like the spiral arrangement of leaves characteristic of many plants, which [in his opinion] has no significance whatsoever in the struggle for existence and is not, therefore, subjected to natural selection.

A similar interpretation of this remark has been adopted by Weismann [33] and Darwin. [Weismann's] is the first attempt to show that changes in the organism due to natural selection must be limited by the nature of the organism itself, i.e. by the chemical and physical nature of its body. He argues from this "that species represent the result of natural selection and the changing potential of their parents". Darwin basically agrees with this interpretation, acknowledging at least that some acquired features can become very stable, even though they play no part in the struggle precisely because of these reasons. He assumes, for instance, that voice, feathers and thorax all have the same significance for the organism, which is why natural selection may pick any of them, but we know that their systematic significance may be different. If a duck-billed platypus had had feathers rather than hair, it would probably have been placed among the birds, not mammals. In such cases, according to Darwin, the whole question is reduced solely to those "qualities of mutability"[16] of the parents which are responsible for the inheritance of various stable characteristics, even those that are not very useful or frankly useless. We shall not analyze this theory further, since our present purpose is merely to show that not only specific and familial features may be insignificant but also that distinctions between larger groups may also be independent of natural selection.

So all scientists, including, by the way, Darwin himself, gradually came to the conclusion that natural selection is not the only factor that changes species and explains their origin. The question which now seems to arise quite logically is the extent of this factor [i.e. natural selection] and the capability of other factors to influence the origin of species. All the previous studies reported in this text may serve as a partial answer to this question, since it has been shown that natural selection, which cannot produce trivial distinctions in humans and lower animals, must not be considered as a factor of primary significance in the production of distinctive specific features.

REFERENCES

[1] One can foresee, *a priori*, that the result of such an observation would not support the theory of natural selection. It is known that oceanic islands each have a relatively small number of species, yet competition between species is most intense precisely because space is so limited. The same applies to aquatic plants floating at the surface; in spite of frequent overpopulation, they are not notable for their specific diversity.
[2] *Archives des Sciences physiques et naturelles*. 1870, p. 160.
[3] Cf. Darwin, Charles. *The Origin of Species*, vol. 1, p. 308.
[4] Darwin, Charles. *The Descent of Man* . New York: D. Appleton and Company. 1871. Vol. I, Ch. IV, (p. 154).
[5] *Ibid.*, p. 248.
[6] *Ibid.*
[7] *Ibid.*, p. 249.
[8] *Ibid.*, Vol. II, p. 354.
[9] *Ibid.*, p. 360.
[10] *Ibid.*, pp. 366–367.
[11] *Zeitschrift fur Ethnologie*, 1871, S.133–143, and 349–359.
[12] *Antropologie der Naturvolker*. I, p. 56.
[13] Die Grundlage der Ethnologie in den heiographischen Provinzen. *Zeitschrift fur Ethnologie*, 1873, memg. VI, S. 322.
[14] Pechsel. *Volkerkunde*, S. 521.
[15] *Antropologie der Naturvolker*, II, S.511 and the following.
[16] *Mitteilungen der Vereius fur Erkunde zu Leipzig*, 1873, S. 25.
[17] Muller. *Algemeine Ethnographie*, S. 428.
[18] One should spread to man an opinion expressed by Darwin with respect to animals on page 316 of the first volume.
[19] Darwin, Charles. *The Descent of Man*, vol. II, p. 365.
[20] *Ibid.*, p. 366.
[21] *Ibid.*, pp. 354–355.
[22] *Ibid.*, p. 355.
[23] *Ibid.*
[24] *Ibid.*, pp. 314, 315.
[25] *Ibid.*, vol. II, p. 352.
[26] Peschel. *Volkerkunde*, S. 72.
[27] On inner differences between Negro and Europeans, see memories of Bruner Bey on Negroes. Memoires de la societe d'Anthropologie,
[28] *Zeitschift fur Ethnologie*, 1874, III.
[29] *Algemeine Ethnographie*, S. 49.
[30] Both Darwin and Wallace agree that natural selection affects mental and ethical characteristics of humans, but neither of them is able to prove the same with regard to physical characteristics of races.
[31] Wagner, R. Handwöztezbuch dez Physiologie, vol. IV, SS. 717, 788.
[32] *Entstenung und Begriff des naturhistorishen Arten*, 1865, S. 26.
[33] Weismann. *Ueber die Berechtigung der Darwinschen Theorie*. 1868, S. 27.

NOTES

1 Darwin was certainly not anti-Lamarckian. He regarded natural selection as "the main but not exclusive means of modification" of evolving organisms, and included very favorable references to Lamarckism in the last edition of *The Origin of Species*. Also Lamarckism survived long after Darwin and Wallace came on the scene.

2 Metchnikoff does not give here a precise reference to the Darwin's *On the Origin of Species*; this quotation is from p. 3 of the original edition. (Facsimile of the First Edition. Cambridge: Harvard University Press, 1964). [editors]

[3] From the last edition of *The Origin*.
[4] In all editions of *The Origin*.
[5] T.H. Huxley. 1861. Papers: Man and the apes. C.Vogt. 1864. Book: *Lectures on Man*.
[6] *The Descent of Man*,1st ed. 1871 (2nd (revised) ed. 1874), rather than *The Origin*, 1st ed. 1859.
[7] Or: the addition of consanguinity to the established notion of [anatomical] affinity.
[8] Neither Wallace nor Metchnikoff knew the extent of unused human brain cells.
[9] Literally fourhanded, the Quadrumana was an order of mammals comprising Primates other than man. Metchnikoff obviously used the term, but "primates", "other primates" or "non-human primates" is substituted where it occurs again.
[10] the mistle thrush.
[11] the song thrush (= *Turdus philomelos*).
[12] *The Origin*, Ch. 3, Struggle for Existence.
[13] The numbers of British song thrushes and mistle thrushes have continued to fluctuate. English song thrushes are relatively scarce at present, but no increase in the number of mistle thrushes has been recorded.
[14] Today, *Rattus norvegicus*. In the text, the modern convention for generic and specific names (capitals, italics, punctuation) has been followed. In Metchnikoff's time, however, there were no Laws of Zoological Nomenclature, and perhaps we should not alter his format.
[15] Today, *Rattus rattus*.
[16] Mutable but difficult to eradicate.

CHAPTER 4

ESSAY ON QUESTIONS ABOUT THE ORIGIN OF SPECIES (1876)

I THEORETICAL VIEWS IN EARLIER CENTURIES - BONNET, LINNÉ, BUFFON, PALLAS, LACÉPÈDE - PHILOSOPHERS

The history of natural science shows that periods of domination of theoretical zoology have been replaced by times when scientists, satiated by generalizations stretched to their limits, turned to the analysis of immediate facts. In doing this, they abruptly exchanged one extreme for another and never achieved the proper harmony of inductive and deductive approaches. Now we enter another period of theoretical domination, and, regardless of the significance of apparent scientific progress, so far as experience and actual results are concerned, science still cannot be regarded as having reached a genuine scientific method.

Although our major task is to become familiar with the content and character of contemporary theoretical activity in science, we have to make a number of preliminary remarks about earlier theoretical developments. This is important both for understanding the history of the development of contemporary ideas and for assessing the period, which is certainly outmoded but still lingers on the scene.

The 17th century is the period when biological sciences finally became independent. It was the epoch of the invention of the microscope and its subsequent utilization in empirical research, and of establishing several academies (among them such important ones as the Royal Society and the Paris Académie des Sciences). A number of outstanding scientists, such as Harvey, Leewenhoek, Swammerdam, Malpighi, Wallace and John Ray,[1] made significant contributions to the very foundations of biology. Physiology, microscopical anatomy, invertebrate anatomy, and comparative anatomy were all created in this century, as were two other fields of special importance for us, namely systematic zoology and *naturphilosophie*.

The 17th century, as is well known, had its "Linné", the Englishman John Ray. He was the first to talk about "species" in the sense in which it is used in modern systematics; he also defined the major groups that comprise the animal kingdom and proposed independent classifications for each of these groups.

[.....] Ray, however, in his systematic and scientific descriptions, did not display a genuine scientific outlook. An amazing difference is presented by his contemporary, Leibniz, who extended his philosophical outlook to the whole of nature and, as a result, touched upon major biological issues. Leibniz was not strictly a naturalist but instead became one of the founders of a supra-naturalistic philosophy; nevertheless his ideas should be mentioned here, principally because of his influence on future philosophical developments in biology, including a recent one. Leibniz regarded the "law of constancy" or "continuity" as one of the major theses of his philosophy; according to which nature does not take leaps, and everything in it, physical events as well as ethical phenomena, are interconnected through gradual transition.

[.....] This broad philosophical concept must certainly have made an impression on naturalists, especially as it was shown, a few years after Leibniz's death, that a polyp, previously considered a true plant, could move and digest food as animals do.

[.....] The most remarkable naturalist of the previous century in this respect was Bonnet [.....] His general treatise on nature, published in a short version in 1762 and in full in 1765, illustrates an integral and clearly scientific development of the principle of continuity borrowed from Leibniz.

[.....] From Bonnet, who combined the qualities of a witty and careful observer with those of a thinker, we now turn to the views of two famous 18th century naturalists who profoundly influenced the entire development of the science of life.

The prime position belongs to Linné, regarded by many as the founder of descriptive zoology and botany. Not much caring about deep philosophical analysis, he aimed to encourage the study of organisms, and he applied all his diligence to the construction of a system [of classification] of the animal and plant kingdoms. He invented binomial nomenclature, still in use, and gave us brief, but at the same time concise and clear, descriptions of organisms. [.....] Although 18th century naturalists recognized the bias of the Linnean attitude (Pallas regarded him as "obsessed with a system too scholastically narrowed"), his approach has been preserved as more suitable for research and teaching than the others available. [.....] One should not think, however, that Linné always followed his initial principles. In his later works he did not consider species to be invariable and stable. He suggested, very tentatively, that "perhaps all the species of one family originally constituted just one species", and they then divided into many by the generation of fertile hybrids. He thus contradicted his own principle: "nullae species novae", and, in fact, deleted it from later additions of his *Systema Naturae* (starting with the tenth). He even made the suggestion that new species discovered after Turnefor (first half of the18th century) developed exclusively through interbreeding between old species.

[.....] The theoretical views of Buffon also changed significantly. [.....] Several times in his works he touched upon the question of species and the origin of animals. [.....] Initially he attributed the various racial modifications of humans to the influence of climate. [.....] But, in reality, neither climate nor food can have a decisive influence; the variability of wild animals depends mostly upon another factor "in proportion to the number of individuals, both those that produce and those that are produced." [.....] As a result of his considerations, Buffon formulated the following: "By comparing all mammals[2] with one another and assigning each of them to its particular species, we can see that the two hundred described species can be reduced to a relatively small number of families, or major sources from which, as it seems, all the others originated." (p. 375). [.....] The wisdom and accuracy of many of Buffon's ideas certainly cannot be left unnoticed; in fact, one of his deductions, as will be shown later, is completely identical with a view expressed by Wallace. Another of Buffon's ideas, that the geographical distribution of mammals is somehow related to their historical origin, can be regarded as the first outline of scientific zoogeography.

[.....] Since Pallas's *Memoir on the Variability of Animals* contains not only a critical analysis of the views of Linné and Buffon but [also] some original ideas of the famous naturalist-traveler, I consider it appropriate to review these ideas, so influential in the subsequent history of biology.

Even by making all possible concessions, Pallas could not agree that climate and other external factors, together with increased fertility and therefore numerous offspring, could have such a powerful impact on the variability of organisms as Buffon argued. [.....] Thus, if the species is a constant and unchangeable unit, then what is the meaning of all the variations noticeable in wild nature and, even more, in the world of domesticated animals and plants? Pallas answered this question in the following manner: "Deviations accumulated over the generations from the initial type of a species, which nevertheless do not threaten the unity of the species, constitute, strictly speaking, natural varieties, [which are] far less common than the inconstancies of domesticated species."

[.....] Pallas was even more determined than Bonnet to make it clear that organisms cannot be placed in some sort of continuous line but must be arranged in the form of a tree, the root of which is made up of the lowest organisms, while two trunks represent the animal and plant kingdoms. Each of these respective trunks ramifies further, and Pallas analyzed some of their branches. Thus Pallas expressed very clearly the idea of a link between different parts of the organic world, although in the form of a general plan, not relationship, as is now regarded.

This sketch of the ideas of three major representatives of biological science of the previous century should persuade us that the question of species and their variability was discussed very seriously at that time. [.....] We learn that the idea of a variety or a race, as an incarnation of variability, has been present since the birth of the concept of species; we see also that the structure of animals, viewed as the style or general design of their constitution, as well as a branching system schematically illustrating mutual relationships of organisms, became a frequent topic of discussion in the 18th century; and, finally, we become aware that these topics were discussed independently of the problems of the origin of species.

Now we shall show how these ideas have been developed further, and also to what extent they have become engrafted and spread in science and the intellectual world.

The greatest development and spread of these ideas can be attributed to a Linnean viewpoint. Simplified studies of animal forms became the object of activities of many scientists. [.....]

Erasmus Darwin, grandfather of Charles, was also one of the scientists who considered the origin of species, and in 1794 he published his *Zoonomia*. In this work, which is a generalized essay on the physiology and pathology of the animal organism, E. Darwin tried to apply the Linnean view of the creation of vegetable forms to the animal kingdom. [.....]

After this short sketch of major 18th-century scientific ideas on species, I now turn to an account of purely philosophical views on this matter. In the course of this we shall constantly encounter statements which cannot sustain even the most superficial scientific criticism, but that is not important here. What is really important is the need to express these philosophical constructions to give rational explanations of some general questions about the entire organic world, and humans in particular. While scientists may be satisfied by simply collecting facts, placing them in general categories, and making apparently obvious conclusions, philosophers, by contrast, concentrate their attention on wider issues, important for creating new outlooks.

The French geologist du Maleaux is representative of such thinkers in a quite general way. Although his occupation was in scientific geology, his reasoning on the origin and evolution of organisms turned out to be so completely unsustainable scientifically that it cannot be compared in any way with that of Buffon, Lacépède and others. [.....] On previous pages we have seen [some] quite risky theories of naturalists, but none went so far in his fantasies as du Maleaux. [.....] His contemporaries were well aware of this; Pallas, for instance, compared du Maleaux's theories to Ovidian metamorphoses.

Robine, another author mentioned by Katrfash as a forerunner of Darwin, went even further into the field of fantasy than du Maleaux. Starting from Leibniz's principle of gradual change, he pushed it to extreme limits, and thus was quite graphic in demonstrating the danger of an unbalanced approach. [.....]

Vastly more significant for us is *System de la Nature*, usually attributed to Golbach, since it familiarizes us with the influential philosophical trends of the last century. This book, which caused a sensation when published in 1770, was either the work of a circle of people gathered round Baron Golbach or [others] under the influence of this circle. Among friends of Golbach, I can mention Diderote, d'Alambert, Turgo, Condylyak, Rousseau, and Buffon, who quite possibly influenced that part of the *Systeme* that is the subject of our present interest. In Volume I, Chapter 6, where the physical and moral nature of humans is debated, the question of the origin of humans is also discussed. "Have humans always been around? Were humans always similar to us? What would they be like? Would a species of unknown origin also be eternal? Are they indestructible or are they transient, like indivisible primordial? Were humans always as they are now or were they moving along an endless line of gradual change to reach the contemporary state?" etc. Numerous postulations of *System de la Nature* can be regarded as answers pertinent to these questions; a theory of gradual origin of organisms is thus clearly included here. [.....]

If from the materialism of the last century we now turn to Kant, the major representative of German idealism, we shall see that, with all their differences in philosophical attitudes, both schools were quite close to each other with respect to the problem discussed here. Having considered the idea of the spontaneous origin of living creatures and change from one type of organism to another, Kant added: "*a priori,* from a position of reason, there is no contradiction in such a statement. Observation, however, does not provide any example of this sort." [1]

As evidence for the continuity of living forma, Kant listed "the similarity of many animal species classified in the same general scheme, founded not only on the skeleton but [also] on the distribution of other partsThis analogy of forms, which, despite their differences, seem to have been produced according to some general prototype, reinforces the hypothesis of a general kinship between them, resulting from their origin from one primeval mother through the gradual approximation of one species to others, ... from humans to polyps", etc. Although Kant nowhere took this idea as established truth, treating it as merely a daring suggestion, [2] he was consistent in considering it possible to apply this suggestion to humans.

In his *Anthropologie,* [3] considering one particular change in the history of humans, he assumed the existence of an era in which humans originated from orang-utangs or chimpanzees as a result of some changes in the organs of these latter [species].

Lessing also clung to a view on the gradual origin of humans from some lower animal form, and he suggested that, at the beginning, they had less than five senses. [4]

II. SKETCH OF THEORETICAL IDEAS IN FRANCE IN PRESENT CENTURY – LAMARCK – ÉTIENNE J.S.-HILAIRE – CUVIER – ISIDOR[3] J.S.-HILAIRE

The beginning of the present century is an important epoch in the development of the theory of the origin of species. Lamarck's *Systeme des Animaux Sans Vertebres*, published in 1801, introduced the theoretical views of this scientist on the general and gradual origin of organisms, later developed in more detail in *Philosophie Zoologique* (1809) and in the Introduction to *Histoire Naturelle des Animaux Sans Vertebres* (1815).

Lamarck was a friend and disciple of Buffon, so it is no wonder that this latter influenced his scientific development. [.....] Like Buffon, he accepted the principle of immutability of species, and only later, after intensive studies and reflections, became an adherent and major proponent of the theory of variability and [its part in] the general origin of species.

Lamarck founded his theory on a purely mechanistic view of the organic world. [.....] This mechanistic outlook, together with his belief in gradual geological changes, resulted in the concept of continuity in the evolution of organisms and their consanguinity [descent from a common ancestor]. After meticulous zoological studies of lower animals, Lamarck concluded that species do not remain permanently within the limits of their present definitions, as was widely accepted by contemporary science. He realized that varieties or races do not differ significantly from species, and concluded the "species originate from varieties". Thus, both *a priori* deductions from general principles and inductive generalizations of facts confirmed Lamarckian ideas on variability and the origin of species in general.

These ideas became a theory, presented by Lamarck as a general overview and tested mainly by deductive means. Here are the most important points of his theory. Living creatures originated through spontaneous generation, and this is still occurring. [.....] More complex organisms developed from these creatures later, according to a law of progressive development. [.....] The basis of this phenomenon is "the life force (*le pouvoir de la vie*), which leads to increased complexity of organization and thus exemplifies progress itself" within the animal kingdom. [5] But this alone cannot explain all observed phenomena. [.....] The animal organism is thus affected by the action both of this major force, *i.e.* progressive development, and a secondary or "modifying cause (*cause modifiante*) which interrupts or deflects the activity of the major force". [6] The modifying influence of the environment, *i.e.* climate, is somehow linked to these deflecting factors. [Changes in] climate and new environments lead to new requirements which stimulate the appearance of new organs, and this is the most important part of the Lamarckian theory of continuity. Buffon also considered environmental influences as the cause of changes in organisms which migrate [permanently], but according to him such influences have direct effect: they either enlarge or diminish the organism, as well as changing skin or hair color, etc. For Lamarck, however, the environment as such has no direct influence: it changes behavior and through that creates new organs. [.....] The core of this theory is the observation, confirmed by daily experience, that organs can be enlarged by increased exercise and decreased by insufficient usage. The powerful effect of this phenomenon on organisms is not questioned, but it is no longer considered as an omnipotent and primary factor. [.....]

At the very end of his *Philosophie Zoologique,* Lamarck included a synoptic table of genealogical linkage of the major groups in the animal kingdom, a kind of genealogical tree, like those so often now drawn by German ultraDarwinians. We would rather not go into the details of this scheme, but point out that, among the divergent branches, Lamarck placed a convergence of branches: ungulate mammals, according to his scheme, resulted from a fusion of two branches, [namely] fishes and pinnipedes. This is important as a test of the admissibility of the Lamarckian principle of convergence (which, until now, has presented a problem for the German ultraDarwinians.)

These are the major features of the Lamarckian theory, a truly remarkable theory but one that has failed to produce any significant impact in its field. [.....] There is a striking dissonance between Lamarck's highly abstract and deductive theory and his practical activities: he was regarded as the "French Linné". It is quite natural that such an ambivalent attitude displayed by the author himself seriously diminished the influence of his theory. An unbiased scientific appraisal of this theory seems to be possible only now, when theoretical explorations have become so important and when Lamarck's deductions have acquired much more value than his systematic treatise on invertebrates. His vigorous acclamation of environmental *adaptation* of animals and *hereditary* transfer of acquired characteristics is wholly accepted now. As for two other factors which, according to Lamarck, determine the variability of species, one of these, namely increased usage of organs, still has some support today, although much less significant, while the importance of the second factor, the force of progressive development (or, alternatively, *pouvoir de la vie*) is completely dismissed by some scientists (like Darwin, for instance) and highly praised by others. We shall return to the Lamarckian theory on more than one occasion, but now let me turn to another proponent of the theory of variability of species: Etienne J. Saint-Hilaire.

Although this scientist came forward with his ideas much later than Lamarck (namely in 1828–1837), he cannot be considered as his successor because of a striking difference in their methods and opinions. Etienne Saint-Hilaire is rather a follower and disciple of Buffon, since environment plays an important role in his theory. According to Saint-Hilaire, however, environmental influence is evident not in adults but in embryos, because they are subjected to many more changes. It is also important that these changes can occur in a shorter period of time, not as a slow accumulation of insignificant alterations, but instantly, as a radical break in development. This way, one can explain the absence of transitional stages between species and the abruptness of distinctions between fossil and contemporary species. É. Saint-Hilaire collected substantial data to support his theory. [.....] He turned out to be not only more scientific in his methods than Lamarck but also more careful. For example, he refused to speculate on spontaneous generation and the [first] appearance of organisms as such, and he did not try to answer questions on the general origin of species, *i.e.* on the prototype of all animal forms. He limited the scope of his consideration of the variability of species to the transition of fossil species to contemporary forms; he also tried to show how animals of one species can produce the first representatives of another, related species, but did not go any further. He formulated his thoughts very carefully, and never presented them as ultimate truth. It is interesting that, although his hypothesis is much further from Darwin's than Lamarck's, his methods and general approach resemble Darwin's much more closely than Lamarck's. No wonder that his theory was much more popular at the time than Lamarck's. Cuvier, who virtually ignored Lamarck, took É. Saint-Hilaire so seriously that he had a public debate with him. Because this critical approach to the theory of variability of species by such a renowned scientist as Cuvier had and has [such] an enormous influence on naturalists, we need to analyze it in more detail.

Cuvier's research opened new scientific perspectives in three directions: by observations on worms which initiated a study of lower animals, continued later by some of his disciples; by founding a new comparative anatomy; and, finally, by a new, rational approach to the investigation of fossils. Working in these three fields, Cuvier enriched science by enormous experimental material and made some extremely important generalizations. Proceeding from the assumption that there should be some basic features for all animals, he divided the entire animal kingdom into four major groups (types, in contemporary usage). He also deduced, from a comparison of the anatomy of a number of living and fossil organisms, that certain organs relate to each other, and thus by knowing some one can get ideas about others. [.....]

It is easy to see from the foregoing that all Cuvier's generalizations were largely inductive. [.....] He did not tolerate deduction and thus theory in science. A goal of [theoretical science] is to find a natural system, such as a classification, that reflects the real affiliation of organisms to each other, as it is in nature. [.....] Another objective is to relate any living creature or organ to its environment. [.....] With this in mind, the comparative method should be applied to the greatest possible extent. [.....]

It is known that Cuvier, as a scientist, was completely different from Lamarck and É. Saint-Hilaire. While these latter demonstrated the power of deductive thinking, composing daring theories by applying just pure logic, and tackling the most difficult problems of natural science, Cuvier tried to avoid these problems and to concentrate his efforts on comparative empirical research. [.....] Cuvier went so far as to deny that philosophy had any significance for scientific research. [.....]

From all the foregoing, it is easy to predict Cuvier's view on the origin and meaning of species. He did not really pay attention to this question, and seemed to consider it unsuitable for scientific discussion. [.....] Like Pallas, he regarded varieties as something completely different from species, and took the opinion that the origin of one species from another was a hypothesis refuted by facts. [.....] Cuvier concluded that the theory of variability of species and the descent of contemporary organisms from fossils could never agree with the facts. [.....]

Cuvier's hostility to the theory of variability of species not only resulted in his disciples following unproductive paths but it deterred [other] scientists from constructing comprehensive theories on the origin of species and became extremely harmful. [.....] The dogma of constancy of species became once again (after Linné) universally accepted knowledge; now, however, it was more convincing since it was based on the idea of varieties as totally unimportant modifications that could not alter the properties of species, [which are] constant and unchangeable. [.....]

But although it was influential, Cuvier's stance could not completely erase the ideas of his opponents. Apparently the attempts to find a compromise by É. Saint-Hilaire's son, Isidor, were quite remarkable. With this in mind, he introduced a theory of *limited variability of species*. According to this theory, species are unchangeable only if the environment is stable. Changes in the latter are reflected in alterations to the properties of species. Such new features result from two forces: one, the force for change, is the new environment *per se*, and the other, the force for conservation, is a mechanism for transmission of hereditary properties. Wild animals must change merely because of overproduction, which forces them to occupy new territories. As for domestic animals, according to this theory they are regarded as changeable anyway, even without environmental influence. Thus the theory of limited variability was, in fact, a theory of variability, even though it did not consider the primary origin of all animals from stated ancestors or transition from one species to another. [.....] It is important to note, however, that Isidor Saint-Hilaire was reluctant to admit this. [.....] So, some scientists, like Darwin, consider him as a transformist, while others, like Katrfash, see him as an opponent to transformism. [.....]

Thus the very concept of variability of species, which appeared and flourished in France, became suppressed and practically eliminated there. Only among botanists (Naudin, 1852; Lecoq, 1854[4]), not influenced by Cuvier, can one find some adherents to transformism. [.....]

III. THEORETICAL VIEWS IN GERMANY FROM BEGINNING OF CENTURY UNTIL END OF 1850s – GOETHE – TREVIRANUS – *NATURPHILOSOPHIE* – BAER AND THE POSITIVE SCHOOL – JOHANNES MÜLLER AND HIS SCHOOL – PHILOSOPHERS SCHOPENHAUER AND BÜCHNER

When Kant supported the theory of variability at the end of the 18th century, he mentioned that this theory had many adherents among German naturalists at that time. [.....] Goethe, in his last work, finished only a few days before his death, points out the profound influence of French ideas in Germany, which apparently led to the popularity of the theory of variability. [.....]

Because of world-wide interest in this remarkable personality, Goethe's contributions to science became known to many primarily as a result of his philosophical reasoning. Searching for general morphological types, he deduced that all parts of a flower are merely modifications of leaves, just as the bones of the scalp are much modified derivatives of typical vertebrae. [.....] It is quite important to clarify how Goethe understood this type of organization and whether his understanding was close to the contemporary scientific view. [.....] Although he returned to the problem over and over again, [.....] he never mentioned [the possibility] that the general [animal] type could have been inherited from a common ancestor of all animals. Goethe's interpretation of the general type is certainly close to Bonnet's concept of a general plan of organization, which was also discussed by Pallas and others. The theoretical views of Goethe present no solid evidence that he was an adherent of transformism, *i.e.* the theory of a common origin for animals and plants. [.....]

When the Darwinists Haeckel and Seindlitz suggested that Goethe discussed the factual origin of all higher animals from a common ancestor, this clearly contradicts Goethe's view of the type in his ideal scheme: a distinctive "general design" that nature had in mind at the start of the creation of organisms. [.....] Goethe is certainly one of the most remarkable advocates of the concept of a general ideal type and the comparative method of analysis of organic forms, but this is still very far removed from the theory of successional origin of species. [.....]

[At the time] when Lamarck was introducing the rudiments of his theory to French scientists (sometime before 1809, when his major tractate was published) a Bremen scientist, Treviranus, published the three volumes of his *Biologie*, in which he argued for a common origin for all animal species. [.....]

We can see an interesting development of questions under investigation in the legacy of the Berlin professor, Johannes Müller. At the end of his *Handbuch für Physiologie*, Müller devoted a chapter to the question of varieties in animals and humans. He insisted that races should be distinguished from species. [.....] and that there is not the slightest chance that one species can be produced from another. [.....] He was also quite insistent that spontaneous generation is impossible in nature, just as it never happened in his experiments.

It is interesting to note that, while Müller and his predecessors Pallas and Cuvier, because of their adherence to the positive method, deny [the possibility of] changes and transformations in species, they accept, because of this very method, some cases of creation, as well as the existence of some ideal plan of creative activity. This looks as if, rebelling against deductive theories, constructed *a priori* but nevertheless quite logical, they are eager to return to even less substantiated but widespread traditional views. [.....]

Because of Müller's and Cuvier's influence, science in Germany turned to previously unknown fields of biology, and completely ignored general questions, including the theory of transformation. [.....]

One should not ignore, however, that during the entire period of domination of "positive" science, even in its most skeptical periods, there were still some scientists who advocated the theory of transformation, albeit in a low and indecisive voice. [.....] I shall not list the names and cite the works of those who do not share the common negative attitude to variability of species. The reader can find them in the second edition of Seindlitz's *Die Darwinische Theorie.* [.....]

Strict limitations in the positive approach can be effective only in science, not in philosophy. Schopenhauer [.....] is quite open about the possibility of transformation of species. [.....] His suggestion that the major diagnostic features of animals are of no immediate importance for their survival and occur purely as a result of the transformation of one form into another seems to be of particular significance. This idea later became a foundation of rational morphology.

Another German philosopher who further developed the concept of transfomism is Büchner. His *Kraft und Stoff* can be regarded as a 19th century manual of philosophical materialism, just as *System de la Nature* was for the 18th century. [.....] Generally speaking, Büchner does not propose anything new with regard to transformism, but he is important as one of the bravest and, at the same time, very sober advocate of this theory when naturalists were extremely skeptical of it.

IV. PATRICK MATTHEW, A FOUNDER OF THE THEORY OF NATURAL SELECTION – PHILOSOPHERS – REVIEW OF FIRST TWO ESSAYS BY WALLACE

In his historical sketch on the origin of transformism as a theory, Charles Darwin refers to a number of Englishmen who favored this theory prior to him. The most remarkable of these was Patrick Matthew, who already in 1831 had developed the foundations of the theory of natural selection by pointing out that species change due to the survival of certain individuals with features beneficial for their own survival as well as that of their offspring. Unfortunately this idea lacked sufficient evidence and disappeared without a trace. [.....]

The same basic idea was developed by [other] English philosophers. [.....] In 1852, the famous philosopher Herbert Spencer came to the conclusion that species change under the influence of the environment. [.....] Because of his preoccupation with philosophical matters, his reasoning had no serious influence on scientific research. Science, as we have just seen, did not pay any attention to the astute observations of Patrick Matthew; so what were the chances of recognition for young Wallace and his brilliant ideas? [.....]

In his first essay, *On the Law which has regulated the Introduction of New Species*,[5] Wallace argued the need to recognize the successive origin of species as a prerequisite for solving the major problems of zoology, zoogeography and paleontology. His point was that geographical and geological proximity between neighboring groups is a result of consanguinity. [.....] In this essay, Wallace also canvassed for granting the concept of transformation the status of a scientific theory and listed a large number of facts it can explain. He did not, however, analyze how this transformation occurs. This became the object of his second essay, in which he argued that the theory of transformation as presented by Linné is not sufficient to explain the origin of species; he proposed instead another version of the theory which limits the variability of species to conservation of characters most beneficial for the adaptation of the organism to its environment.

Wallace criticized the current understanding of variation as something very unstable; this assumption was based on the observation that domesticated animals released into the wild often revert to their prototypes. His conclusion was that, on the contrary, species can deviate from their prototypes without limitation and permanently if they acquire characters beneficial for adaptation. To illustrate the full significance of these benefits, Wallace drew attention to the struggle for existence in nature and to the laws of species survival. [.....] These laws can be applied not only to individuals but to species and races as well. [.....] His principal suggestion, that races with certain beneficial features can survive their prototypes, explains not only the origin of new species but also the apparent facts of domestic animals reverting to their wild state. [.....]

This whole brilliant essay is simply deductive development. [.....] As to empirical verification, Wallace supplied this in his essay on Malayan butterflies, published six years later (in 1864), so we shall turn to this essay in one of our subsequent chapters.

V. GENERAL OVERVIEW OF DARWINIAN THEORY – ITS MAJOR POINTS – NATURAL SELECTION AND ITS RESULTS: DIVERGENCE OF FEATURES AND PERFECTION OF ORGANISMS – SEXUAL SELECTION – OTHER FACTORS OF VARIABILITY OF SPECIES – APPLICATIONS OF THEORY OF SUCCESSIONAL ORIGIN – ARGUMENTS IN FAVOR OF TRANSFORMATION

We have outlined the major stages of the development of the transformation theory prior to Darwin. We have seen that the principal idea, that of the origin of one species from another, was published in a scientific form a century ago, and that, despite powerful resistance by major scientific authorities, it has survived until now.

We have followed changes in the ideas of advocates of transformism, including their views about the influence of the environment, and we have outlined how the major objectives of this theory were underlined (by Matthew in England and Naudin in France). But all this is not enough to acquire the status of a scientific theory. Even the most brilliant ideas cannot survive without comprehensive scientific analysis, verification of their major *a priori* statements, and demonstration of their relevance in solving many significant scientific problems. These tasks became a life-long preoccupation for Charles Darwin, the great English naturalist who developed, in the course of long travels and prolonged studies, a remarkable versatility and breadth of view. He combined the qualities of an English naturalist-traveler with those of a German scholar, and, thus equipped, wrote the first short account of his theory. The main points of this account turned out to be almost identical to the theories of Matthew, Naudin and especially Wallace. This account was followed by a major treatise entitled *On the Origin of Species by Means of Natural Selection or the Preservation of Favored Races in the Struggle for Life.* Darwin wrote his first essay [on this subject] as early as 1844, but published it only fourteen years later, when Wallace's text was printed. Immediately after that, in November 1859, Darwin published the first edition of his book, which, since then, has engendered five[6] more editions and has been translated into the majority of European languages. [.....]

Before cataloguing the major Darwinian ideas, it seems quite instructive to follow the way in which they developed. Darwin gave the following explanation:

When I visited, during the voyage of H.M.S. *Beagle*, the Galapagos Archipelago, situated in the Pacific Ocean about 500 miles from the shore of South America, I found myself surrounded by peculiar species of birds, reptiles, and plants, existing nowhere in the world. Yet they nearly all bore an American stamp. In the song of the mocking-thrush, in the harsh cry of the carrion-hawk, in the great candlestick-like opuntias, I clearly perceived the neighborhood of America, though the islands were separated by so many miles of ocean from the mainland, and differed much from it in their geological constitution and climate. Still more surprising was the fact that most inhabitants of each separate island in this small archipelago were specifically different, though most closely related to each other. The archipelago, with its innumerable craters and bare streams of lava, appeared to be of recent origin; and thus I fancied myself brought near to the very act of creation. I often asked myself how these many peculiar animals and plants had been produced: the simplest answer seemed to be that the inhabitants of the several islands had descended from each other, undergoing modification in the course of their descent; and that all the inhabitants of the archipelago had descended from those of the nearest land, namely, America, whence colonists would naturally have been derived. But it long remained to me an inexplicable problem how the necessary degree of modification could have been effected, and it would have thus remained for ever, had I not studied domestic productions, and thus acquired a just idea of the power of Selection. As soon as I had fully realized this idea, I saw, on reading Malthus on Population, that Natural Selection was the inevitable result of the rapid increase of all organic beings; for I was prepared to appreciate the struggle for existence by having long studied the habits of animals. [7]

Even a superficial knowledge of domesticated animals and plants evidences a significant degree of variability. Although it is still not completely clear why this happens, environmental changes are responsible for such variations in many cases. Sometimes these changes affect animals immediately, sometimes their influence is indirect, through breeding. Among the most important rules of direct environmental influence that one can mention is the law of analogous changes, when alteration of one organ leads to changes in certain others. [.....] Even more obvious seems to be the hereditary transmission of acquired features. Not only innate but some acquired characters, like certain diseases, become hereditary. [.....]

Variability of features and their hereditary transmission are two major factors which facilitate the creation of so many domesticated animals and plants, each for its own purpose. [.....] People learned how to do this long ago: they selected for breeding only those individuals which could be useful for human purposes; that is why they chose stronger or healthier individuals, or those with special features relevant for certain purposes (like sense of smell, quality of hair, etc.). It was also important to isolate these chosen individuals from the rest of the group to avoid loss of the desired features through interbreeding. [.....]

Darwin extrapolated these rules of domestication to the whole of organic nature. Although domesticated animals and plants are certainly more prone to variation, wild animals and plants also change, sometimes quite drastically. [.....] If it is true that variability is inherent to organisms that are not directly influenced by humans, then it seems quite likely that heritable variability is a feature common to all organisms. The question now is whether there is something in wild nature comparable to the domesticating selection performed by humans. [.....] There are no factors in wild nature which improve or preserve the breed, but food shortages are always to be found, triggered both by natural disasters (flood, cold, etc.) and over-reproduction. Malthusian law is certainly applicable, not only to humans but also to the [entire] animal and plant kingdoms. [.....]

Nature, according to Darwin, is always overpopulated, and this leads to a struggle for existence between individuals for food, shelter and procreation. This struggle, as in any competition, is strongest between individuals with identical needs which relate to each other in the most intimate manner. So, this competition is more intense between members of the same species than between neighboring species, and so on. It is quite obvious that the winner will be the individual with certain favorable features.

[.....] This process of survival of the fittest and supplanting of the weakest is called by Darwin *natural selection*. Like artificial selection, which preserves individuals useful for human goals and eliminates the useless ones, natural selection singles out individuals better adapted to the struggle for existence and destroys those less fortunate. [.....]

Because the complexity of the environment complicates the struggle for existence, natural selection works in many different ways; so it can frequently happen that a feature selected in one case is destroyed in another. [.....] But are there some preferred paths of natural selection, in spite of all these numerous choices? Darwin points out two such paths. [.....] The first results in divergence of characters. Natural selection in general favors variability, because it facilitates the very survival of organisms. The more any two forms differ, the greater their chances of occupying different natural niches and therefore of having a better "survival prospect". [.....] The same law of divergence of attributes constantly manifests itself in human life, as in a situation of fierce competition with too many applicants seeking a certain position, individuals choose other occupations. [.....]

Since the production of individual features is extremely widespread in nature, divergence of characters, the basis of this phenomenon, is certainly one of the most important aspects of natural selection. Darwin was accused of paying too much attention to divergence and underestimating the converse, convergence of characters. It is quite easy to come to the *a priori* conclusion that if, for instance, two neighboring species produce a large number of variations because of the law of divergence, extreme variations may converge and thus produce a single new species. In spite of the apparent veracity of this suggestion, Darwin was still convinced that complete convergence is never possible and, therefore, the principle of convergence must be rejected.

According to Darwin, specialization or perfection of organization is another feature of natural selection. Interpreting this generally as specialization of organs for certain functions, *i.e.* division of labor between organs, Darwin considered this a very beneficial acquisition in the struggle for existence, and thus necessarily embraced by natural selection. [.....] This provided him with a way of explaining progress in the animal and plant kingdoms over geological time, a phenomenon that Lamarck attributed to a hypothetical striving for perfection.

Accepting that, in general, natural selection improves qualities in the majority of cases, Darwin did not apply a perfectionist attitude to all situations. [.....] Indeed, in parasites, for which the environment is quite simple, the opposite change, *i.e.* regression or primitivization of the organism often takes place. This can also explain why some forms do not perfect their organization for a long time. Changes would not be beneficial for these forms and therefore [such changes] are likely to be eliminated by natural selection. [.....]

Natural selection [.....] picks out features which are important for survival not only for individuals but also for entire species. [.....] Thus natural selection is joined by sexual selection, which picks out features ignored by the former. Sexual selection deals with features facilitating procreation, not the crucial struggle for existence. [.....] Darwin considered this form of selection as very important. Sexual selection, according to him, is responsible for many differences between males and females, including color, voice, general outlook, and ornamentation. [.....] In short, for all the so-called secondary sexual features, indirectly related to breeding. [.....]

In the same way that natural selection picks features important in the struggle for existence, sexual selection singles out peculiarities significant in the sexual struggle, or competition between males for the possession of females. [.....]

Features acquired by males through sexual selection are inherited usually but not always by their male offspring; sometimes these features are transmitted to both sexes. Such cases are extremely interesting in connection with the origin of species, since they can explain the appearance of features apparently useless for natural selection. [.....] According to Darwin, this gives a clue to the origin of racial differences in humans.

Natural selection can work differently in gregarious animals, which often acquire features beneficial for the entire species but neutral or even harmful for some of its members. [.....]

Darwin considered natural selection (together with sexual selection) to be "one of the most important factors in the modification of living forms", but not the only force behind the origin of species. There are some cases of modification which cannot be explained by either natural selection or sexual selection. He cited the rule of correlation of parts, according to which some organs depend on each other to such an extent that, if one is changed by natural selection, the other will change too, even if there is no immediate benefit to the organism as a whole. This correlation is often observed in so-called homological organs, like the anterior and posterior extremities. [.....]

The utilization of organs, according to Darwin, can also play some role in their modification. This factor, which was recognized long before Darwin and served as a cornerstone of the Lamarckian theory, was greatly valued by Darwin and used to explain cases in which natural selection does not seem to be working.[7] [.....]

According to Darwin, the environment exerts an immediate and significant influence. This position is drastically different from that of [earlier] transformists. [.....] As evidence for his view, Darwin referred to species which can live in a range of environmental conditions. He also pointed out that internal changes in organisms are much more likely than those resulting from external environmental influences. [.....] In some cases, however, external influences can be crucial, especially the reactions of some plants (mostly flowers) to changes in color. [.....]

Thus Darwin did not claim that natural selection could explain everything about the variability of species, but he was certain that selection is the main factor in the origin of species. Darwinism, therefore, can be tested only through cases that may be explained by natural selection, or by the law of correlation [of parts], or the law of utilization [of organs], or, finally, by the direct influence of the environment.

Darwin [not only] successfully explained a large number of facts completely unfathomable to previous science, [..... like] perfection of organization and different forms of specialization, [.....] geographical distribution of organisms, [.....] morphological or systematic relationship (which he reduced to "community of descent") [.....], but he countered earlier objections to any attempt to explain the origin of species from other species. A detailed study of factual examples convinced him that a strict division between species and varieties is not possible; on the contrary, the latter should be regarded as an incipient version of the former, and all differences between them are relative and quantitative. According to Darwin, the absence of transitional stages between forms, a strong argument against transformism, can be explained by their rapid extinction: they are always weaker than the more distinct species, and thus have no chance in the struggle for existence. The absence of transitional forms between fossils cannot be proved due to incompleteness of the geological data.

The stability of certain forms, such as those depicted in ancient Egyptian manuscripts (an argument that was so important for Cuvier and other [critics of] transformism), was of no significance to Darwin.[8] Natural selection does not aim to change species at all costs; it often acts in the opposite direction, removing changes and preserving major forms. [.....]

VI. SKETCH OF POSTDARWINIAN THEORIES ON TRANSFORMISM – THEORY OF HETEROGENETIC[9] ORIGIN. – THEORY OF PROGRESSIVE DEVELOPMENT AND THEORY OF DIRECTIONAL CHANGE – THEORY OF SUDDEN CHANGE – THEORY OF MIGRATION AND DISSOCIATION

The theory, the foundations of which we have just presented, spread in England and all over Europe at unprecedented speed. One reason for such success was that minds were already prepared to accept new teaching, since Cuvier's and Müller's theories, certainly powerful enough to slow down the development of transformism, were not able to stop it. Scientists who were afraid to express their opinions during the domination of the positive school, now openly joined the transformists. Thus, because of the conclusiveness and comprehensiveness of *The Origin of Species*, in a short time transformism penetrated all branches of biological science. [.....]

But while the general transformation theory was developing rapidly, the particular form called Darwinism or the theory of natural selection was experiencing serious difficulties. There were frequent attempts to explain the variability of species by factors other than natural selection. [.....]

First among these new theories of transformism was the theory of heterogenetic origin (*Theorie der heterogene Zeitung*) developed by Kölliker, a well-known specialist in microscopic anatomy and animal morphology. A major element of this theory is that animals change because of some inner causes which may sometimes be undetected, as when humans in general grow older and look older, but any particular individual may die before biological old age and thus never look old. Species keep the capacity to change, but do not always go through the actual changes. [.....] The actual change, according to Kölliker, occurs through rapid and sudden transformations. This claim was extremely important for Kölliker because it makes transitional forms unnecessary; their absence was always a stumbling block for transformists. Kölliker attempted to explain the sudden creation of new species by the phenomenon of transformation (metamorphosis) in insects and asexual reproduction in many lower animals. In this latter case, one animal can bud off another animal which can look quite different, as in the case of a hydropolyp and a medusa. [.....] Like É.J. Saint-Hilaire, clearly his forerunner, Kölliker associated all modifications of species with their embryonic stages only. To justify this opinion he cites the well-known fact[10] that the major organs of all animals are formed very early, even in the embryonic stage. [.....]

It is important to note that the whole theory of heterogenetic origin was derived almost entirely by deduction. [Limited] data were used merely as foundation for far-reaching and often unrealistic analogies, so it is no wonder that only some ideas of this theory have been accepted. [.....] The most valuable of them is the idea of an inner changing ability inherent in the organism. Here Kölliker was close to another transformist, Nägeli, Professor of Botany in Münich. [.....] Nägeli defined this capability as acting predominantly toward increased complexity of organization, [....] according to a fixed plan. [.....] Such a postulation reflected his inability to explain all facts of complexity solely by natural selection. [.....] From this position, however, the very existence of lower forms should certainly be regarded as paradoxical. [.....] To get round this difficulty, Nägeli turned to the hypothesis of spontaneous generation, and explained the production of lower forms from unicellular organisms by analogous transformations which occurred in different geological periods. [.....]

There is no need to analyze this theory in detail to see that its foundations are very shaky. To support this theory, other suggestions, even more far-reaching and contradictory, have been put forward. Many facts testify directly against this theory, like the occurrence of species unchanged over several geological periods. [.....]

Nevertheless, Nägeli's theory was generally accepted by German botanists, who regarded natural selection as capable of explaining some facts only, while the most crucial development, namely the creation of formative, systematically important characteristics, cannot be explained other than by a specific inner drive of organisms to change. Among those who shared this theory was Ashkenazi,[11] who came to the conclusion that variability [brings about] progressive development, *i.e.* increased complexity of organisms. [.....]

Making the unquestionable statement that physiologically important adaptations are not necessarily morphological, Ashkenazi still could not refute two major tenets of Darwinism, namely that natural selection leads to perfection and to divergence of characteristics. Both Nägeli and Ashkenazi did not even attempt to clarify the correlation between natural selection and variability. [.....]

Among zoologists who supported the theory of progressive evolution and determinate variability (orthogenesis) was Weismann, who came to this conclusion independently. [.....] He favored the application of this theory to the animal kingdom, reasoning that natural selection is not sufficiently powerful to explain the transition to more complex animals. According to him, natural selection tends to have the opposite effect, namely that of simplification of organization. He shared the opinion of the anonymous author of a pamphlet published in Hannover in 1872 (*On the Blending of Species by Means of Natural Selection*), [claiming] that selection of organs is far less beneficial in the struggle for existence than blending, and that highly specialized organisms are at a disadvantage compared to those more adapted to the environment. [.....] Another antiDarwinian argument is that natural selection would lead to simplification of organisms because simple organisms are more common and therefore more fit. [.....] With these arguments, he tried to ridicule the theory of natural selection by adducing gradual simplification of organization under natural selection, finally transforming humans to apes and higher animals to lower ones. To avoid such capricious conclusions, Weismann proposed a principle of evolution with limited variation and a qualified type of natural selection. [.....]

Without analyzing these arguments now, let us note in passing that they are not as serious as Weismann perceived them to be. Natural selection, according to Darwin, frequently but not necessarily leads to a complexity of organization. [.....] There are some cases of so-called retrogressive development when natural selection simplifies a complex organization. [.....] In general, this discussion shows how a one-sided and exaggerated presentation of the original Darwinian arguments can become a caricature of Darwinism.

It is easy to see, therefore, that a new school of transformists, who admit the importance of natural selection but consider an inner drive to mainly progressive development to be much more important, has been formed in Germany. This drive is responsible for [the view that] morphological features form the foundation for the classification of organisms and are thus crucially important for systematics, while natural selection influences only physiological characteristics significant in the struggle for existence. The contributions of this theory are mainly negative, relating to the just-mentioned limitation of natural selection to changes in physiology, not formative features, and to its emphasis on the nature of variability. [.....] Its positive contribution is not very impressive, since it fails to show that variation occurs in only one, progressive direction. [.....]

Many transformists argue against the use of natural selection to explain the initial formation of organs, as in the case of protective colors and forms which can appear quite rapidly. [.....]

Anticipating the criticism that the creation of organs by a slow selective process is not always the case, Darwin suggested that some organs that become useless or even harmful, as a result of changes in the environment, can acquire new functions. These acquired functions are performed by organs already developed by natural selection, so there is no need to slowly accumulate alterations for these organs to show a drastic change in function. [.....]

While the transformists of this "progressive" orientation were substantiating their position, relying mostly on variability underestimated by Darwin, Moritz Wagner, another dissenting transformist, emphasized [the importance of] isolation of selected evolving individuals. [.....] Darwin considered isolation to be a major element in natural selection, *i.e.* survival of the fittest, but, according to Wagner, natural selection alone is not really sufficient to prevent crosses between "the fittest" and weak individuals already defeated in the struggle for existence. [.....] When Wagner was reminded that complete geographical isolation, certainly important for the creation of new species, had never actually been observed and recorded, he set aside natural selection and replaced it with his "separation theory" (*Separationtheorie*). [.....]

This theory was subjected to criticism by Prof. Weismann , [.....] who cited cases of side-by-side existence of several closely related species, as well as long-term separation not resulting in new species or races. While not exaggerating the significance of separation, Weismann considered it quite important for changes in species. After careful analysis of mechanisms of separation, he came to the conclusion that periods of change in the duration of any species would be followed by times of stability, and that the creation of a new form can take place only if isolation occurs in a period of change. [.....]

Transformists of all sorts thus proposed major innovations to explain the question of successive origin of species. It is quite clear from the above that natural selection alone cannot be sufficient. Not only the transformists but Darwin himself never argued otherwise. The present objective is to show the limits of natural selection and the importance of other factors in the formation of species. This incredibly difficult and important task is still far from resolution, [.....] but guidelines for it are under intense scrutiny. Once they have been established, a comprehensive study of data relevant to the successive origin of species will finally become possible.

What remains for me now, therefore, is to outline the major factors in the creation of species, familiarize the reader with available data, and, finally, indicate all possible fields and applications for future generations of naturalists.

VII. CHANGES IN DOMESTICATED SPECIES – CASES OF SUDDEN APPEARANCE OF RACIAL FEATURES – ON DISTINCTIONS BETWEEN DOMESTICATED AND WILD ANIMALS – FERTILITY AND INFERTILITY IN HYBRIDS – APPLICATION OF VARIABILITY OF DOMESTICATED ANIMALS TO QUESTION OF SUDDEN APPEARANCE OF RACIAL FEATURES

Transformists have always been aware of changes in domesticated animals and plants, which are often rapid and drastic. [.....] Their opponents, however, also found major arguments in the field of domestication. A comparative study of wild and domesticated animals demonstrated a drastic difference between species and races. While transformists stress the striking variability of domesticated animals, where nearby breeds often differ more drastically than species or even families in wild nature, their opponents decry the significance of such facts by referring to some well-known rules of animal breeding, namely that interbreeding of domestic animals usually results in fertile offspring while [distinct] species are unable to interbreed or the hybrids are not fertile.[12] [.....] It is clear from this that Darwin was not in a position to ignore domestication, and indeed he faced this question in his brilliant book *The Variation of Animals and Plants under Domestication.* As for Wallace, he certainly underestimated the significance of domestication; his conclusion that domesticated animals are abnormal, being subjected to changes that never occur in the wild, is biased and not sufficiently substantiated. [.....] This divergence of opinion between the two great founders of the theory of natural selection is quite revealing when the prevalence of Darwin is considered. [.....]

In the following review I shall rely almost entirely on the above-mentioned book by Darwin, with no fear of delusion by any bias he may have had. The conscientiousness of this great naturalist is as famous as his genius and comprehensive erudition.

The very fact of significant variability in domesticated animals is well known, so there is no need to argue about it. It is sufficient to mention here that not only breeds descended from more than one species but also those from a single ancestral species are subject to such variation. In this latter case, variability is not to be explained by interbreeding, it is inherent.

Such natural, inherent variability is the only way in which new races could have been created. [.....] Darwin presented the general conclusions of his studies on domestication in the very last chapter of his treatise:

"Some breeds have been from their first origin modified in so slow and insensible a manner, that if we could see their early progenitors we should hardly be able to say when or how the breed first arose; whilst other breeds have originated from a strongly-marked or semi-monstrous deviation of structure, which, however, may subsequently have been augmented by selection. From what we know of the history of the racehorse, greyhound, gamecock, &c., and from their general appearance, we may feel nearly confident that they were formed by a slow process of improvement: and with the carrier-pigeon, as well as with other pigeons we know that this was the case. On the other hand, it is certain that the ancon and mauchamp breeds of sheep, and almost certain that the niata cattle, turnspit and pur-dogs, jumper and frizzled fowls, short-faced tumbler pigeons, hook-billed ducks, &c., and with plants of a multitude of varieties, suddenly appeared in nearly the same state as we now see them." [8]

In the first of Darwin's categories, certain racial features have been enhanced by quantitative growth, while qualitative change is most important in the second category, which is marked by the sudden appearance of new forms. [.....]

The next question is, how applicable are these conclusions from domesticated animals to those not dependent on humans? Actual variation in wild animals is unquestionable, but its frequency and degree are far less known and, according to Darwin, depend on environmental differences. These vary more in the case of domesticated animals; also [the supply of] food is better when supplied by humans, which presents an additional source of variability. [.....] Since natural selection is practically absent in domestic animals, all modifications, both beneficial and harmful, are retained, while in nature only beneficial changes survive; this too promotes variability in the former. [.....]

Variations, however, also occur in wild nature, so it is quite rational to ask the aforementioned question. To answer, we should first distinguish species from varieties or races. Unable to define morphological differences, naturalists turned to physiology, and claimed that species do not cross, or that crossing produces fertile offspring only in exceptional cases. This claim, advanced mostly by non-transformists, became decisive in the definition of a species. [.....] Transformists have attempted to refute this assertion by [citing] recorded observations on the breeding of hybrids produced by two different species. [.....] These observations, however, indicate only that there must be some exceptions to this claim regarding infertile progeny, but it does not undermine the claim itself. Darwin presented another interpretation of this entire phenomenon which was much deeper and more straightforward.

He accepted that hybrids are generally infertile and searched for a reason by comparing the fertility of hybrids with that of other animals transferred to different environments. [.....] Darwin compared the profound changes in sex life due to environmental transfers to those which occur in crossing different species or their hybrids. [.....] In both cases the reasons are still unknown, but it is well established that the sexual system is affected first and foremost by various changes. [.....]

Darwin presented evidence that crossing different forms within the same species can, in some cases, give fully fertile offspring while in others fertility is either reduced or totally absent. Hybrids from completely fertile crossings are very similar to those between different species. [.....] This shows the primary importance of sexual fertility for hybrids but not for species or races. [.....]

Thus one of the major arguments of the nontransformists, that of a profound difference between species and races, is rejected. This permits far-reaching analogies between species and races. Careful consideration must now be given to their differences in fertility. Of the two ways of changing domestic animals, by the accumulation of individual differences and through the sudden appearance of racial differences, Darwin picks only the first and applies it to nature. He states that, in the vast majority of cases, wild races and species are produced by the accumulation of individual differences governed by natural selection. The sudden appearance of new species in the wild is considered by him to be highly unlikely. [.....] His major argument is that, even if such a sudden change occurs, crossing with unchanged individuals would be inevitable. [.....]

Darwin, discussing the possibility of sudden changes, does not mention natural selection, for as yet unknown reasons.[13] If this factor is able to select individual alterations, why cannot it do the same for more drastic, sudden changes? This question was not raised by Darwin, so there is no direct answer; but we may try to answer him using his own data. [.....] My response seems to contradict his claim that sudden changes cannot survive breeding. On the contrary, such deviations have a better chance of lasting than small ones.[14] [.....]

A third objection, or rather remark, of Darwin was that the majority of features that appear suddenly are simple cases of return to the ancestral condition. [.....] Although this remark was clearly an *a priori* statement and not a rule, it can be accepted without changing Darwin's major argument. An individual that re-acquires some ancestral feature will already differ from its ancestor in many respects; natural selection then has the task of conserving or eliminating this feature. [.....]

The study of domestic animals can thus be linked to the theory of successive origin of species and, in fact, strongly supports it.

VIII. STRUGGLE FOR EXISTENCE – DARWIN'S VIEWS ON STRUGGLE – MAJOR MOMENTS IN STRUGGLE FOR EXISTENCE – THEIR ROLE IN ORIGIN OF SPECIES – INFLUENCE OF STRUGGLE FOR EXISTENCE ON MORPHOLOGY OF LIVING CREATURES – EXTINCTION OF LOSERS IN STRUGGLE, AND NäGELI'S STUDIES ON THIS QUESTION

The very notion of a struggle for existence was introduced into science long ago, primarily by botanists. They noticed the supplanting of some plants by others, a process which was often compared to forcing out primitive species by European ones, particularly as the supplanters were mainly of European origin. Thus the concept of fitness of a plant in the struggle for existence appeared. [.....]

On the subject of the "struggle" in the plant kingdom, botanists have used this idea in a restricted sense [.....], usually referring to two plants in the same immediate location. [.....] They have not made any connection between this phenomenon and the variability and origin of species. [.....] The idea of a link between them comes from the founders of the theory of natural selection. Both Darwin and Wallace considered the struggle for existence as a complicated process resulting either in survival or death. [.....] According to them, the struggle for existence also includes competition between individuals with identical needs [.....] and thus is an extremely complex phenomenon affecting the most diverse individuals.

Both Darwin and Wallace saw unrestricted reproduction of animals and plants as a major cause of this struggle. They extended the Malthusian view on humans to the entire planet Earth, which is overpopulated and where each life depends on several deaths. In order to survive, a living creature has to win its life or die.

According to Darwin, the struggle for existence, especially where the origin of species is concerned, is governed by the following major law: the struggle is at its most intense among individuals and varieties of the same species. This is so because neighbors have to compete for the same food, living space, etc. [.....] Darwin thus gives priority to competition between forms with identical needs, especially individuals and varieties of the same species, which leads to the origin of new species. Some transformists, like Nägeli, have recently reduced the entire struggle for existence to this competition. One can go even further and claim that the struggle for existence is applicable only to competition between individuals of the same species, because competition between species cannot lead to the origin of a new form. [.....]

Such a restriction is intolerable. [.....] I intend to show that the struggle for existence consists of several important components, with different mechanisms. These components are: 1) competition between individuals of the same species; 2) competition between individuals of different species; 3) the struggle between individuals of different species, as between predators and herbivores, and 4) the struggle between living creatures and natural disasters (cold, drought, etc.)

Since, from the Darwinian position, species are created mainly through the accumulation of individual or racial differences beneficial in the struggle, it is quite obvious that struggles in the first category are of prime and immediate importance for the creation of new species. The others are significant only as stimuli for competition between individuals of the same species, like, for example, the struggle between predators and their prey which intensifies competition both within the predators and their victims. Winners could thus be found everywhere: not only the predator able to get the biggest catch but also the hunted that develops the most effective individual disguise.

In addition to the three kinds of stimuli mentioned, unrestricted reproduction is considered by Darwinists as one of the major factors triggering competition between individuals of the same species, and leading, in the long run, to the origin of new species. This factor is especially important as the only one generating competition when other stimuli are absent. [.....] This helps Darwinists to explain how the first species on earth was able to split without any external competition. [.....]

How do these four stimuli correlate with each other? In some places in *The Origin of Species*, especially when discussing freshwater plants and the population of isolated oceanic islands, Malthus's law, one of the foundations of Darwinism, seems to be ignored. Darwin's argument, that freshwater plants occupy relatively small areas and therefore competition between them should be less intense, illustrates this. The same argument is made with respect to the population of oceanic islands. The facts and their interpretation are indisputable: overpopulation, impossible under such conditions, does not lead to the creation of new species.

These two examples indeed confirm that unrestricted reproduction and overpopulation by one particular species are far less important than mutual struggle between different species populating the same territory.[15] This might help to avoid some contradictions found in *The Origin of Species*. [.....] It is quite natural to assume that simple overpopulation of one species is not powerful enough to trigger the mechanism of natural selection; it requires the competition of as many species as possible. It may be invalid to say that natural selection originates only when there is an assortment of species, but it is fair to claim that primitive conditions and scarcity of primeval populations are not conducive to natural selection.

Let us now turn to the question of the most important characteristics of organisms in the struggle for existence. One can claim, as did Nägeli, that such characteristics embrace everything from extending longevity to more intensive breeding. Such suggestions, however, contradict major principles of Darwinism. Indeed, if heightened reproduction leads to a struggle for existence, how can this be reconciled with the tenet that the same quality greatly facilitates victory in the same struggle? [.....] In spite of Darwin's initial view, it is clear from my arguments that increased fertility is not nearly as important for triggering a struggle for existence as competition and conflict between different forms. However, when competition between different forms is taking place, fecundity becomes very important, and the winners will almost certainly be the most fertile. The reverse result might occur when the struggle for existence is generated by overbreeding: less fertile individuals have an advantage because numbers are less important than endurance and strength. [.....]

It is certainly quite important to trace the link between fecundity and morphology, i.e. to investigate whether similar forms have the same fertility or whether this feature does not depend on morphology. There are insufficient data to formulate a definite answer, so we can only speculate. [.....] There is no clear evidence that systematic relationship of species reflects a constant level of fertility. [.....] *Morphological features, therefore, are not directly indicative of a level of fertility and cannot act as a criterion of success in the struggle for existence if, in this particular case, the struggle is caused by high fertility.*

Another important factor in the struggle for existence is endurance or strength. This is not merely physical strength, but the ability of one creature to force out others, regardless of whether or not they are physically stronger. [.....] Darwin seems to have [expressed] no definite opinion on this. He argued theoretically that "endurance" reinforces the adaptation of the organism to the environment, and he considered this as one of the crucial points of his entire outlook. Practical checks, however, depend on knowing why some organisms are better adapted and thus supplant others. He was very uncertain and pessimistically admitted, in the chapter on "Struggle for

Existence", that probably in no such case are we able to decide why one creature wins and another loses. [.....]

Botanists who have specifically studied this phenomenon with regard to plants have come to a similar conclusion. They have emphasized two aspects: "intermittence of victory" and lack of "obvious reason" for it. All this indicates is that survival is not determined by morphological distinctions between competing species but by something deeper and less obvious. Many botanists, like A. P. de Candole, think that occupation of the same area for a long time can make a plant weaker. [.....] This would explain not only the periodical replacement of some plants by others within a limited area, but also supplanting of native flora of oceanic islands [.....] by "newcomers" of more recent origin. This supplanting, however, is often not complete because of seeds of the old plants remaining in the soil; they start to grow when the "newcomers" ultimately weaken due to the same reason of isolation. Darwin never openly refuted this opinion, but he obviously did not support it. [.....]

Regardless of this latter discord, the major conclusion of both Darwin and de Candole is still valid: namely that *survival does not depend on morphological characteristics of species but on certain physiological peculiarities; features vital in the struggle for existence and systematic (morphological) specific characters do not necessarily coincide.*

We can make the *a priori* assumption that frequency of environmental change is an important factor in winning the struggle. [.....] There is no need to list here all the characteristics of organisms that can be beneficial in the struggle for existence and, at the same time, utterly unimportant in their morphology. [.....]

In my essay on *Anthropology and Darwinism,* I have already explored this point of view (on the absence of any regular connection between features beneficial in the struggle for existence and morphology) with regard to humans, and I referred to Nägeli, according to whom beneficial adaptations in this particular case are, without exception, physiological. [.....]

There are winners and losers in the struggle for existence. The latter (and this is one of the major points of Darwinism) die so quickly that gaps often occur in the sequence of related species. Such early extinction is beneficial for the winners because it isolates them and they are prevented from interbreeding with the losers. [.....]

This proposition by Darwin has been challenged by Nägeli, who, for more than ten years, studied the question of the origin of species from varieties of the same species. His conclusion is that analogous competing forms do not produce winners that force out losers (in the usual sense). Rivals in the same area tolerate each other, and competition between them determines only their relative abundance and never leads to extinction. This statement undermines Darwin's notion of the complete extinction of losers. Competition without the possibility of extinction, however, hardly seems compatible with the origin of new species under the Darwinian scheme, as well as many [cases of] the absence of intermediate forms between [related] species. [.....]

This conclusion is certainly disturbing and potentially important for framing a satisfactory theory on the origin of species. It seems to be reflected in Nägeli's speculation that "social life is more important than isolation for the creation of species".

This also demostrates how little we understand some of the governing factors in the struggle for existence or any struggle, and how slowly we have progressed in this field since the initial conceptualization by Darwin and Wallace. To illustrate this latter suggestion, I draw the reader's attention to the minimal changes in the relevant chapter of the latest edition of *The Origin of Species.*

IX. FOUNDATIONS OF THEORY OF NATURAL SELECTION – REVIEW OF FACTUAL EVIDENCE – FACTS UNEXPLAINED BY SELECTION – TENDENCY TO IDENTICAL VARIABILITY – LAW OF DIVERGENCE OF FEATURES NOT COMPLETELY REDUCIBLE TO ACTIVITY OF SELECTION – PERFECTION OF ORGANIZATION AND THEORY OF SELECTION – RELATIVE IMPROVEMENT – PROGRESS, REGRESSION AND CONSERVATION IN ORGANIC WORLD

It is quite obvious that winners in the struggle for existence have to be those more adapted and more adaptable, *i.e.* creatures with features advantageous in this environment which are absent or not sufficiently developed in the losers. [.....] This foundation of the natural selection theory is so well established that nobody doubts the very existence of natural selection and the survival of the fittest. What needs further clarification is the significance of this struggle for the creation of organic forms, as well as its actual mechanisms. This point is also reinforced by the range of opinions held by different transformists on this subject. Darwin considered natural selection as the main, but not the only, cause of the origin and modification of species. Nägeli, on the other hand, came to the conclusion the natural selection in the Darwinian sense does not occur in the vegetable kingdom. Between these extremes there are many intermediate positions, like that of Weismann (discussed in Chapter VI.) [.....]

Irrespective of the interest and importance of Darwin's books published after *The Origin of Species*, they do not fully clarify major points of his theory. In addition, Darwin's condensed style is not always unambiguous, and this leads to misunderstand and differences of interpretation (especially concerning divergence of features). [.....] But there is still no better source than Darwin's books, and we turn to them now. After presenting major *a priori* principles of natural selection (in Chapter IV of *Origin of Species*), Darwin indicates that features that are quite unimportant morphologically can become subjects for selection. For example, camouflaging hair color in some animals can change with environmental changes; this is crucial to individual survival and hence the survival of the species. [.....] This confirms the final deduction of my previous chapter, that, according to Darwin, features of the utmost importance for survival, and therefore picked out by natural selection, do not necessarily coincide with the range of features of morphological importance. [.....]

With the notable exception of humans (*The Descent of Man*), Darwin never attempted to apply the theory of natural selection to any particular species. Such an attempt was made by Wallace with regard to the Papilionidae,[16] spectacular butterflies of the Malayan archipelago. His conclusion is that the evolution of Mayan Papilionidae conforms to the theory of natural selection especially as far as striking forms and colors are concerned, but it is not completely proved that natural selection can explain all or even a significant part of the morphological characteristics of the Papilionidae.

Writings of the founders of the theory of natural selection indicate 1) that cases of obvious and visible natural selection are helped by "strong heredity", 2) that natural selection often works on hidden features, less apparent than the morphological peculiarities of races and species, and 3) that changes in these latter cannot always be attributed to natural selection. [.....]

In his most recent publication, Darwin turns to mutually beneficial cooperation between some flowering plants and insects. These mutual benefits are attained through various systems of complex relationships picked out by natural selection in the most appropriate way. [.....] But the question still remains: how significant, from the morphological point of view, are changes initiated by such selective cooperation, i.e. are they sufficiently far-reaching to be reflected in systematics? I shall try to answer this question using data of the German scientist Müller,[17] clearly a Darwinian and a very conscientious scholar. His studies clearly indicate that both insects and flowering plants do not show any significant variability in their morphological organization. There might be an objection [to this view] if the flowers of flowering plants and the organs of digestion in insects had no systematic importance, but the facts are otherwise. [.....]

I attempted to show on previous pages that the main evidence for the very existence of natural selection testifies to the inability of this factor alone to explain the relationships of organisms. Now I shall approach the same target using Darwin's theoretical ideas on natural selection.

Natural selection, according to Darwin, should be able to alter organisms of all ages, accumulate beneficial changes, and transmit these age-related alterations to future generations. [.....] Considering this together with a major idea of Darwinism, namely that natural selection is triggered by overpopulation, one can argue that the younger the organism, the stronger the selection should be. But if we compare different groups of the same species, of different ages, living in a certain area, it is obvious that the youngest part of the population is always the most numerous. Mortality, however, is also highest among these youngest individuals, even in humans, which clearly reflects an intensified struggle for existence here.

From these facts and the just-cited major idea of Darwinism, one may conclude that morphological distinctions should be most marked in embryos, less in larvae, and still less in adults. The facts, however, do not support this conclusion, either with regard to embryos or to larvae. (This is particularly so in the case of larvae, since those of different species often look surprisingly similar.[18])

How then are we to explain specific differences between adult organisms that greatly exceed those between their embryos and larvae? Without rejecting the theory of selection, we can suggest that the characteristics of adult forms are directly related to, or even depend on, those features of embryos and larvae that resulted entirely from natural selection. This explanation is indeed [sometimes] applicable, but not in many cases. [.....] Sexual selection, which, according to Darwin, will often explain phenomena not covered by natural selection, is not very helpful here either. Female insects, always less numerous than males, cannot practice sexual selection because their eyesight is inadequate; males, with big eyes adequate for the recognition and thus selection of females (and not used for anything else), do not select either, because there are so many of them.[19] [.....]

So, visual features seem to be not very important where natural selection is concerned. It certainly is not possible to explain this situation now, and it is also still too early to assess the advantages and disadvantages of the theory of natural selection. Such an assessment may become a life-long preoccupation for an entire generation of scientists. Exactly because of this, however, it is useful to indicate those forms whose racial differences are not governed by natural selection, such as Rhizocephala (e.g. *Euglipia*), some insects (e.g. *Seira domestica*), sponges, and even humans. [.....]

Darwin himself admitted the possibility of modification of features independent of natural selection. The only difference between our points of view is that Darwin considered these cases as exceptional and was not really concerned about them, while in reality they are quite common. [.....]

The major effects of natural selection, according to Darwin (see Chapter V of this book), are divergence of characters and improvement of organization. The first effect is particularly important since it shows the benefits of variability. Divergence of characters follows overpopulation as a way to reduce the intensity of struggle and to allow many more individuals to survive. [.....] If divergence is really as important for the creation of new species as Darwin considers, we should expect it to develop at the same time as overpopulation. This, however, does not seem to be the case, because, as we saw in the previous chapter, organisms most often subjected to overpopulation are extremely resistant to morphological change. We also know of cases of invasion, as in the case of oceanic islands, where "newcomers" have been able to remain and prosper in spite of a dense population of native species which was ultimately extinguished. [.....]

Overpopulation, then, is frequently not followed by variation. [.....] Major changes among "islanders" do not follow a divergence principle, which, according to Darwin, should change neighboring forms first to give them a chance to survive through spreading into vacant environmental niches.

The most variable species are the most widespread, as Darwin himself noted. He attributed this fact to environmental differences and competition between different species. Divergence is clearly pushed into the background, particularly as differences in organization appear earlier and independently from divergence. [.....]

So the divergence principle, while clearly important, seems not to be as crucial as the Darwinists believe; otherwise its effects would not be obscured so often.

The second major effect of natural selection, namely improvement in organization or progressive development of organic life, seems to be much more important than the first. [.....]

The very idea of progress in the organic world is so popular among zoologists, botanists and geologists that Darwin considered it his duty to investigate the possible connection between this idea and his theory. He immediately warned the reader that the very idea of progress in the form of animals and plants is unclear, and the only sound criterion is the von Baerian one, [which recognizes] a degree of independence of parts of an adult animal and their specialization for different functions. Darwin thus reduced the principle of progress to [an effect of] natural selection on the grounds that specialization of organs is beneficial for the organism. According to Darwin, this reduction is supported by cases of regressive development in parasites and some other forms. [.....] Darwin was cautiously optimistic about general progress in the organic world, although he considered it difficult to prove.

Some naturalists, prior to Darwin and independently of transformism, like Bronn, attempted to investigate this issue, but almost immediately discovered the Baerian criterion of perfection to be inadequate. [.....] Bronn's conclusions were quite close to those of the philosopher Herbert Spencer, who later and independently stated that progress reveals itself in a series of isolated instances accompanied by their concentration and subsequent merger.

Because of the great importance of this issue, and the influence of Bronn's theory on science, I shall attempt to clarify my own attitude to these two major definitions of progress.

Von Baer had earlier noted that application of his criterion of perfection can lead to some quite paradoxical conclusions. For example, many insects and cephalopods are more perfect than fish, and the skulls of fishes are superior to those of birds, mammals or even humans, etc. To avoid such nonsense, von Baer qualified his criterion with a new principle, the so-called principle of integration, first introduced by Spencer. This theorem is based on the rule of reduction and merger of similar parts, which are more numerous the higher a given organism occurs in systematic classification. [.....] Applying this principle entirely consistently, without arbitrary exceptions, we can eventually reach, at the pinnacle of progress, forms considered to be lower because of retrogressing development, so-called regressive forms. [.....] In order not to contradict accepted scientific knowledge, Bronn and Spencer limited their investigations and stopped at insects, lobsters and the like. They focused on measuring progress relative to the supreme perfection of humans. Since in these latter cases we can see both perfection of organization and simplification, any general formula of progress should necessarily include these. [.....]

In spite of all the definitions of this expressly anthropomorphic scheme, it is not satisfactory in cases where an objective assessment is required, especially at the lower end of ladder of the organic world. [.....] So this approach is not applicable to all classified creatures, and, in addition, it is subjective to a degree that cannot fail to shock naturalists. From a scientific point of view, humans cannot be regarded as the embodiment of supreme perfection in the organic world. A number of essentially mammalian organs are unduly suppressed in favor of the predominant development of the brain. Humans should not have to lose organs like the tail and ear muscles to become perfect. [.....] From a position of objective perfectionism, humans must be regarded as too specialized and thus unbalanced creatures, like the Helicopidae among butterflies. [.....]

Because of this confusion, many naturalists, like Hermann Müller, completely abandoned any attempt to find a criterion of perfection. According to this ultraskeptical approach, progress in the organic world is meaningless, and therefore no correlation between natural selection and perfection of organization is possible. This outlook was certainly unacceptable to Darwin, and, since the Baerian criterion seemed more objective than that of the anthropomorphists, he used this when correlating natural selection with progress, dealing with separate parts, and only later with other minor factors of progress, like merging. [.....]

Darwin's observations show that even he was not in a position to demonstrate a definite correlation between selection and progress. [.....] It is quite obvious that there is not the strong link between them that Darwin presumed. [.....] This can lead to the suggestion (already made by several scientists) that progressive development, unexplained by natural selection, depends on some inner drive to perfection, inherent in all organisms. Before accepting this far-reaching statement, we need [to consider] some additional arguments. It seems obvious that winners in the struggle for existence are helped not only by peculiarities of their own organization (in the most comprehensive sense) but also by the environment. The fittest individual under certain circumstances may be a loser under others. [.....]

It is known that lower organisms are spread more widely in space and also in time. [.....] One may infer that this is so because the more tenacious lower organisms are restricted to simple environments and survive in places not accessible for a more complex and developed life. [.....] Many naturalists believe that, in earlier geological eras, lower organisms were pushed into the background and gradually gave up their place to higher organisms. [.....] On the other hand, there is an opinion (expressed, as we saw, by the anonymous author of the Hannover essay) that natural selection favors lower organisms as more tenacious. But this view is a biased objection to Darwin's idea that natural selection promotes improvement of organization. Perhaps what really happens in nature is something in between: in some cases selection favors higher forms, while in others it does not take progress into account and selects features entirely irrelevant to criteria of perfection. [.....] It is obvious that perfection is mainly associated with specialization of organs [.....], so environmental changes affect higher organisms much more than lower forms. [.....] We may conclude, therefore, that *natural selection favors improvement of organization in a stable environment only; environmental change naturally leads to the selection of intermediate and lower organisms.* This rule can also be applied to restrict one of the important generalizations of Darwinism, namely that intermediate forms are the most commonly eradicated. [.....]

So if, as it seems, natural selection does not favor higher, more advanced forms, does this mean that the rule of progressive development in the organic world, largely accepted by naturalists, depends only partly on natural selection, and thus [also] relies on an inherent striving for perfection? [.....]

It was noticed long ago that higher organisms appeared later than lower ones. [.....] Since the former are more complex, an inference about progress in the organic world seems quite natural. This process is often regarded as directed toward perfection of organization. [.....] Even if we reject this idea as biased and acknowledge that there are many cases of regressive development, natural science maintains some general idea of organic progress. [.....] But this general idea is clearly unsatisfactory without attempting to measure degrees of progress and investigating the reverse process, *i.e.* regression. [...] This latter is far more widespread in nature than is realized.

Parasites provide rich material for such studies, but they are not the only source. Taking the animal kingdom as a whole, we cannot claim that regression is prevalent, but there are clear cases in insects. The Hymenoptera is probably the only group of insects where progress is quite obvious. [.....]

Clear signs of regression can be noted in higher animals including vertebrates (e.g. reptiles[20]), as well as in the vegetable kingdom. [.....]

The facts of regressive development are quite difficult to reconcile with the theories of the progressionists, but, surprisingly, they are more often in accord with natural selection than with progressive development. To construct a general picture of changes affecting organisms, one has to admit that neither progression nor regression are entirely responsible for them. Facts show that the most widespread type of evolution in nature is conservation. When new forms appear it is by the creation of species higher or lower than their relatives. This is noticeable in systematics, which places related forms in a linear sequence. [.....]

In another essay [9] I made the deduction that evolution is a far more complex phenomenon than simply a progression. It also cannot be explained as a sum of differentiations and integrations; it can include transformations and atrophies. In fact, "conservative" species are quite numerous, and the majority cannot be explained by natural selection. [.....]

XI. THEORY OF TRANSFORMISM IN RELATION TO DISTRIBUTION IN SPACE AND TIME – INTERMEDIATE FORMS – TRANSFORMISM IN MORPHOLOGY – POSITIVE AND NEW *NATURPHILOSOPHIE* – OUTLINE OF GASTRAEA THEORY

The geographical distribution of animals provided the first hint to Darwin and Wallace in the construction of their theory. [.....] They thus became transformists even before developing the idea of natural selection. [.....] Almost the same can be said about [their ideas on] the successive appearance of organisms. The majority of Darwin's statements on this point are directly related to transformism and only a few to the theory of natural selection. [.....] It is true that such important phenomena as the extinction of species and improvement of organisms are consistent with natural selection, nevertheless Darwin does not attempt to show that they cannot be correlated with other types of transformism. (The possibility of explaining these phenomena by other transformist theories was demonstrated in previous chapters of this book.) [.....]

The Origin of Species clearly influenced morphology, the science of the constitution and relationship of forms. This effect, however, was not due to natural selection alone but to a revival of the theory of successive origin of species. Hence we can extend to morphology the above remarks on geography and geology, namely that both transformism and Darwinism play their part. [.....]

The "general plan of organization", conceived by comparative anatomists, turns out to be, from a transformist position, the totality of features of a common ancestor. Rudimentary features [.....] were also quite easily explained by Darwin. [.....] He revived the *Naturphilosophie* idea (see Chapter III) on similarity of embryonic development as an indication of common origin.[21] [.....] But since it was already recognized that [adult] morphology has accumulated a significant amount of data on the "general plan of organization" and the relationship between different forms, it became relatively easy for the new transformists to merely place them in a transformist formulae. [.....]

Although numerous examples of similar organization and development of forms which differ in many respects can support transformism, this is not the case for the many related forms with quite different ("acquired later") organs. Morphologists who investigated these cases, like F. Müller and O. Schmidt, came to the conclusion that the homology proposed by previous scientists is only apparent. [.....]

Since the new transformists still use the same methods of analysis, finally changing only the interpretation of the "general plan" to relationship, the search for homology [.....] has been transferred to this new school without any significant change. Major proponents of this trend in comparative anatomy and morphology are Huxley in England and Gegenbauer in Germany, both zealous transformists. [.....] The so-called "most recent naturphilosophers" represent another viewpoint of scientists inclined to comprehensive generalizations. When the former disciple of Johannes Müller, the zoologist Haeckel, started his actual research, he did not show much talent or profoundness, just patience and diligence. He published several books on the lower organisms, but he became famous mostly as a result of his vigorous defense of Darwinism and transformism. Becoming a convert to Darwinism, he abandoned the high research standards of his famous teacher, Müller, and he did not adopt the incomparable scientific merits of his new master. In his *History of Creation*,[22] Haeckel attempted to make Darwinism understandable to the general public and presented it as a "naturphilosophical" system, especially fashionable in Germany. In his effort to expand Darwinism as a system, Haeckel proposed the most daring and unsubstantiated hypotheses, and finally lost track of them and their separation from scientific truth. [......]

Haeckel transferred the same methods, clearly marked by dilettantism, from his popularizing books to his scientific tractates. He published a series of essays in which he attempted to extend Müller's work on the Crustacea to the entire animal kingdom. Haeckel became a supporter of an idea put forward by the first school of *Naturphilosophie* and revived by L. Agassiz as the "law of embryonic types", according to which early organic forms correspond to the embryos of later forms. It was subsumed into his theory by Darwin, and developed further with respect to invertebrates by Müller and to vertebrates by Rutimeyer. Haeckel proclaimed this idea as "the greatest biogenetic law", and defined it as follows: "ontogeny (*i.e.* individual development) is a repetition of phylogeny (i.e. evolution of the species)".

Applying this law to the entire animal kingdom, Haeckel attempted to find, through "ontogeny", the common ancestor to all animals. He started this search when a number of scientists, particularly Russian zoologists headed by Prof. Kovalevsky, devoted themselves to the study of major problems of comparative embryology. At the very beginning of his research, Kovalevsky established that the early stages of development are similar in many animals: a deepening indentation in the ball of cells [blastula], resulting from cleavage of the egg, becomes the primary digestive channel [archenteron] , so that this early embryonic stage already has two concentric sacs, external skin and internal digestive sac, this latter opening at the primary mouth. He also discovered that in some animals this development takes place within the egg capsule, while they are still embryos, while in others it occurs when they are transformed into larvae. The double sac just mentioned is covered with vibrating cilia when it emerges from the egg capsule and starts its free life in the water. [.....] He suggested that this mode of development may be universal throughout the animal kingdom, but he never asserted this, due to the fact that comparative embryology had not provided the necessary data for such a general conclusion. [.....] But, as often happens, when a great scientist is hesitant to draw a crucial conclusion because of insufficient proof, this claim is made by a less careful dilettante. In our case, this role is played by Haeckel with his gastraea theory. Everything really valuable and scientifically proven in this theory belongs to others, mostly to Kovalevsky. Haeckel merely extended the mode of development investigated by the Russian scientist to the entire animal kingdom (except the Infusoria), and set up his own terminology (sometimes quite apt, but mostly useless) and a predominantly scholastic classification.

I do not consider it either possible or appropriate to analyze this theory in detail (this should be done in specialized journals), but I cannot ignore it because of the reverberation it has caused in science.

Haeckel divided the animal kingdom into two large groups: (1) primitive or unicellular animals, and (2) "intestinal" animals, which have tissues and a digestive channel. According to Haeckel, animals of this second group (the Metazoa) are all descended from one ancestor, called "gastraea". This did not survive, but its vestiges can be seen in the larvae of many animals in what Kovalevsky calls the "gastrula", the "double sac" mentioned above. [.....]

Haeckel based the scientific foundation of his theory on [the assumption] that gastrulas are completely identical in all animals studied, and the two layers are homological with those of the Coelenterata. The opinions of scientists differ on the reasons for this morphological correlation. Some say [the layers] are of similar origin, while others consider the similar organs developed from them to be responsible for the correlation. Haeckel is inclined to the latter view, so, to test his theory, he must show the similarity of both layers of the gastrula in all "intestinal" animals.[23] He is sure that this is precisely the case, but such a claim is at least premature, due to lack of sufficient data. If we turn to the origin of the layers, we also find here significant discrepancies with Haeckel's theory. [.....]

All the above is quite sufficient to conclude that, if Haeckel had adhered to accepted scientific method and based his theories on his own careful research (as Kovalevsky and others have done) rather than drawing arbitrary conclusions, he would never have rushed in with his ready-made comprehensive theory. Haeckel regards this theory, together with his other creations, as part of a philosophical outlook on nature which he call monism. Because of this, many of his readers think of Haeckel as the founder of a new philosophy. In reality, however, (as Haeckel himself admits) his monism is nothing more than a mechanistic outlook developed long ago and generally accepted by a previous generation of naturalists. He may merely claim authorship of the name, the converse of dualism, which has been recognized for a long time.

This outline of the viewpoints and results of transformist morphology shows, amongst other things, that the theory of transformism has come to the forefront of the field, and it replaces the previous concept of "general plan of organization". Particular forms of this theory, including Darwinism, are still developing. [.....]

By way of conclusion to this analysis of the origin of species, I should like to draw the reader's attention to the fact that, in spite of the participation of the greatest scientific minds in the exploration of transformism, the search for the origin of species is just beginning. I have very often had to restrict my inquiries because of insufficient data, particularly when the struggle for existence, natural selection, and variability are concerned. It would certainly be difficult to do justice to all these and other questions before a whole new generation of scientists joins the exploration. One of the major objectives of my outline is precisely to encourage this future activity by pointing out relevant questions and by drawing attention to serious studies on transformism.

REFERENCES

[1] Kant. *Kritik der Urteilskraft*, 2nd edition, S. 368–370.
[2] *Ibid.*, S.370. Quite important is Kant's remark that there are not so many naturalists who wouldn't suggest the same about the origin of organic forms.
[3] *Anthropologie*, 2nd edition, 1800, note to p. 323.
[4] Quot. in *Zeitschrift fur Volkerpsychologie*, vol. VIII, 1875, S. 276.
[5] Lamarck. *Natural History of Invertebrates*, vol. 1, p. 134.
[6] *Ibid.*, p. 114.
[7] Darwin, Charles. *The Variation of Animals and Plants under Domestication.* London: John Murray, Albemarle Street, 1868, v. I, pp. 9–10.
[8] *Ibid.*, v. II, p. 414
[9] *Vestnik Evropy*, 1871, N 4, pp. 752–757.

NOTES

1 John Ray, botanist, 1627–1705.
2 The 200 species suggests mammals rather than animals.
3 There are various ways of spelling this name, but Darwin refers to him as Isidore.
4 Spellings from Darwin.
5 Published in 1855.
6 There were six editions in all.
7 Darwin scarcely mentioned the inheritance of acquired characters in the first edition of *The Origin of Species*, but by the last edition he was almost a Lamarckian. Metchnikoff was probably aware of this shift.
8 Darwin's concept of evolutionary time was much wider than that of most of his contemporaries.
9 We prefer "heterogenetic" (from "genetic", pertaining to origin) to "heterogenous" (which could be derived from "gene", a word not yet invented).
10 This "well-known fact" is demonstrably untrue in many phyla.
11 Or Askenazy (*fide* Tauber & Chernyak).
12 Exceptions are considered in subsequent paragraphs.
13 Probably because he had already discussed it in *The Origin of Species.* Darwin repeatedly affirmed that natural selection could work only through gradual changes.
14 This argument about blending of both large and small differences was not resolved until Mendel's work was rediscovered.
15 Darwin's view is supported by the Galapagos finches, believed to have evolved from one species.
16 Swallow-tail butterflies.
17 Fritz Müller.
18 Whereas others look surprisingly different.
19 This applies to hive bees and some other social Hymenoptera, but not insects in general.
20 Presumably Metchnikoff was thinking about the loss of limbs in snakes.
21 Williamson claims that genetic recipes for embryos and larvae are transferable.
22 *The History of Creation* (New York, 1876)
23 The various types of gastrula are stages in metamorphosis from blastulas to various metazoan groups. (Williamson)

CHAPTER 5

COMPARATIVE EMBRYOLOGICAL STUDIES (1881–1885)

COMPARATIVE EMBRYOLOGICAL STUDIES I.
(WITH FIGURE XII)
I. ENDODERM FORMATION IN THE GERYONIDAE[1]

The Geryonidae seem to be particularly useful in considering the formation of endoderm in the Metazoa; the fairly large size of their embryonic cells facilitates such a study. I studied the embryonic devlopment of *Carmarina hastata* as early as 1870, and found that "the blastoderm divides into two layers, with a transparent gelatinous mass between them." [1] A year later, Folle [2] studied the development of *Gyryonia (Carmarina) fungiformis* and observed the same division of the blastula into two embryonic layers. Our data, however, was disputed by Kovalevsky in 1873, [3] who considered that the blastodem in *Carmarina hastata* becomes invaginated; this invagination, according to him, subsequently leads to the formation of endoderm. Kovalevsky observed the same division of blastoderm cells but interpreted it differently from Folle and myself. The following quotation is from Kovalevsky's Russian text:

> Cell division is very rapid and results in the formation of a one-layered blastoderm consisting of fairly elongated cylindrical, or more correctly truncated-pyramidal, cells surrounding a small cleavage cavity.[2] The nuclei are located at the outer ends of the cells, while the inner parts are filled with transparent gelatinous matter. The separation of the outer and inner parts is soon observed; the former is transformed into surface epithelium, while the inner parts fuse with each other to form an almost completely homogeneous, gelatinous mass, which fills up the cleavage cavity to form a gelatinous mass in the young larva; a small invagination in the surface cells at one end of the embryo becomes the larval stomach; the ventral side of the outer layer, close to the mouth, thickens, and small rudiments of nodulose tentacles develop, and the formation of the umbrellar cavity begins.

While some English embryologists, like Ray Lankester and Balfour, accepted the ideas of Folle and myself, Haeckel [4] disputed the delamination which we observed; he is clearly unwilling to acknowledge the very existence of this process as something that results in the formation of endoderm. Haeckel's disciples, Hatscheck and the Hertwig brothers, supported him. The former [5] claimed that "cases of delamination are questionable", and the latter [6] asked how frequently, if at all, examples of gastrulation through delamination occur. Such skepticism can be explained, on the one hand, by Kovalevsky's data, and, on the other, by results described but not sufficiently proved by Folle and me. If we turn to Folle's publication, it is quite apparent that his description of the process of delamination is not as precise and detailed as his other descriptions; the only figure to illustrate this process (Fig. 9 [of Folle's paper]) relates to a fairly advanced stage of endoderm formation and depicts [only] the external contours of the cell. Lack of drawings leads to schemes that cannot be regarded as particularly conclusive. My 1874 results are more reliable in this respect because they were based on drawings of delamination made from life.[7] These results, however, are not complete: they do not cover the internal changes of the formation of endoderm cells, and they lack essential details for the description of such an important phenomenon.

In order to elevate the process of delamination in the Geryonidae to a position which does justice to its overall significance and meets the requirements of comparative embryology, in 1880 I conducted studies on two species of Geryonidae – *Carmarina fungiformis* Haeck. and *Liriope eurybia* Haeck. – at Professor Dohrn's Zoological Station at Naples.[3] On 20 March a female of the first species laid numerous eggs, which soon began dividing. Next morning the process of delamination was well under way in several eggs; this certainly eliminates any doubt about genuine delamination – it certainly took place here. Another observation in seawater gave exactly similar results to those described in my article mentioned above. A drop of dilute acetic acid added to the egg was sufficient to get a much clearer picture of the blastomeres and their structure. Dr. 1 shows an optical section after the fifth division, i.e. the stage with 32 blastomeres. In all eggs observed, the fine grained ectoplasm could easily be distinguished from the generally reticular endoplasm. The majority of the blastomeres were in the process of division (mostly tangentially, but sometimes also radially). The orientation of cleavage is determined predominantly by the nuclear spindle; this is easily seen by comparing eggs a and b (Dr. 1). While radial cleavage leads to the formation of new elements of the blastoderm, tangential division results in separation of ectoplasm from endoplasm or in the appearanceof the primary elements of endoderm.

[The fact] that [cells in] the same egg can divide in many different ways, and that these ways can be effective simultaneously in time and space within this same egg, is an important test of the generally accepted view of embryonic development. The peculiar lacinate form, often acquired by the nucleus after cell division is completed, apparently serves as the foundation for the previously prevailing view on embryonic development.

All blastomeres of the same egg do not divide simultaneously; this creates a certain irregularity and makes a characteristic imprint on each egg.

It is quite difficult to obtain healthy eggs of *Carmarina* through rearing, and much easier in *Liriope eurybia*, a smaller member of the Geryonidae so widespread near Naples. Fertilized eggs of this species can be obtained just a few hours after sexually mature specimens of this medusa are placed in a bowl. I observed such specimens in March, April and May. The only problem is the speed of early developmental processes, which is so rapid that eggs have to be artificially cooled to slow down these processes at night.

Delamination in *Liriope* is basically of the same type as in *Carmarina.* It is important to note that both species are quite easy to obseve. Several blastoderm cells extend into the cleavage cavity [blastocoel]; their inner ends swell like buds. Application of acetic acid immediately results in [the nulei] developing the familiar spindle-shape (Dr. 2). In the middle of the spindle one can see Bütschli's granules, which is entirely consistent with current views on cleavage of egg cells [8]. Optical sections are so clear and convincing that it would be pointless to prepare man-made sections. In order to present a complete picture, however, I have included a diametrical section in Drawing 2, also Drawing 3, which shows several cells from the upper surface in such a way that the delaminating cell "a" is in the center.

After the division of the nucleus is completed, a rearrangement of the protoplasm within the cell takes place; ectplasm remains almost entirely in the peripheral part, while endoplasm, by contrast, moves to the inner part (Dr. 4). The newly-formed nuclei are of different shapes [in the two cells], which earlier served to test the view that the nucleus has a simple structure. It should be mentioned that delamination in *Liriope* does not prevent division of cells at the radial surface to create new ectodermal elements (cells "d" and "e", Dr. 4, are evidence of this). The newly-formed endoderm cells also divide into two (Dr. 4, f).

The completion of delamination is marked by the formation of a complete sphere of endoderm cells, separated from the spherical ectoderm by a formless gelatinous substance. This is the well-known diblastula, so familiar to everybody from the earlier works of Folle and myself.

COMPARATIVE EMBRYOLOGICAL STUDIES II.
II. ON SOME STAGES OF *CUNINA*, A PARASITE OF *CARMARINA CYNOCTANTHA* (HAECK.)

When the inadequacy of Kovalevsky's assertion on the embolic development of endoderm in the Geryonidae became apparent, I had to revise other data on gastrulas of hydromedusae. The first of these data came from studies on *Tubularia mesembrathemum* by Chiaminchian, [9] according to whom the endoderm is formed quite early as a type of so-called epibolic gastrula. A closer look at his work raises doubts on the correctness of his deductions; [10] the development of endoderm in tubularians can be explained in a completely different manner. Both Balfour and Kleinenberg [11] were unsuccessful in confirming Chiaminchian's ideas, and my own efforts also failed. The tubularians, together with many other hydromedusae, first develop a solid morula, which only later splits into two embronic layers. This follows both from my own observations and from the earlier studies of Kleinenberg, who very kindly showed me many of his preparations clearly demonstrating delamination of the morula.

I must reject Chiaminchian's suggestion [12] that *Polyxenia flavescens* (*Aegineta flavescens* or *Salmoneta flavescens* Haeck.) undergoes epibolic gastrulation. By 1870, I had quite often observed the transformation of a fertilized egg into a solid morula, four blastomeres remaining unchanged for a time but later following the others. The morula is thus composed of completely identical cells which only later delaminate into two embryonic layers. Although I was not able to repeat these observations, I found the same solid morula with its subsequent delamination in my study of the earlier stages of development of *Solmessus asbescens* Haeck. (*Cunina albescens* Gegenb.). Cell division and delamination in this medusa are very similar to the corresponding processes in siphonophores.

It is quite clear from all the foregoing that I do not share Haeckel's view [13] on the occurrence of an amphigastrula in *Salmoneta flavescens.* What Haeckel took for an amphigastrula was merely an early stage of morula, for which the time of differentiation of endoderm is a very remote prospect.

Uliyanin, also, thought that hydromedusae have a genuine gastrula. He described a larva of *Cunina*, a parasite of *Carmarina*, as an oval gastrula with a terminal blastopore. Although he did not observe the hatching of this larva, he considered it "certain that one half of the one-layered ball [of cells] in this animal (and also in the majority of others) is sticking into the other half." [14] That is why he considered this larva of *Cunina* to be an *Amphigastrula invaginata*. As a peculiarity of this larva, he mentioned "the filling of the entire cavity, which is limited by endoderm and connected to the external milieu through a hole which penetrates the endoderm, by a fine-grained mass which contains different sorts of vacuoles." Uliyanin searched in vain for this mass in the core of cells, but in some cases he was able to see a fairly big round body. He interpreted the mass as digestive secretion of endoderm cells, while he regarded the round bodies as foreign inclusions, probably the remains of digested food.

The youngest stage that I discovered of parasitic *Cunina* (or *Cunoctantha*, as it was called by Haeckel) was found on the edge of the umbrella of *Carmarina fungiformis*. It can be seen with the naked eye as a small milk-white spot; under the microscope it appears something like a rhizopod with a round cap (Drs. 7 and 8). My immediate impression was that this peculiar larva consists of a gigantic amoeboid cell and a bell-like dome composed of contractile epithelium. The giant cell produces a small number of similar offshoots, many of which branch and develop discoidal expansions. Inside this protoplasmic mass is a large nucleus, quite similar to the central capsule of many radiolarians, together with a thick elastic membrane and a mass with granules and with tubular particles of different sizes (Dr. 9, t). The nucleus is surrounded by numerous fatty corpuscles which, quite surprisingly, never penetrate the pseudopodia. This gigantic cell, which is the dispersal mode of the young larva, is nothing other than the fine-grained mass with round body observed by Uliyanin inside the cavity of the "gastrula". Such a mass, however, is not the unique phenomenon that Uliyanin thought; I saw it in every specimen. In some cases, however, instead of one big nucleus, I found two smaller ones, as seen in Dr. 8. Such exceptional cases were also spotted by Uliyanin. It is not easy to observe the living specimen: although a layer of contractile cells is visible, its contents remain a mystery. This can be clarified only by man-made sections, to achieve which I prepared larvae in Kleinenberg picrocyanic acid and stained them in Grenacher borax carmine. After decoloration in acid alcohol and penetration by ergamot oil, larvae were embedded in paraffin wax and sections were cut. The most important thing seen in the sections is the endoderm (Dr. 3, 9, en), a combination of irregularly shaped cells surrounded by contractile epithelium (ectoderm). There is, apparently, no gradual transition from ectoderm to endoderm; on the contrary, there is obvious penetration of the protoplasm of the large locomotive cell by the protoplasm of [other] cells. This locomotive cell is later visible in section as an extended body without pseudopodia, probably acquiring this shape at the time of death. The same apparently happens to the fatty corpuscles, so that the entire protoplasm is now fine-grained. Although it is probably too early to make an assessment of the true histological significance of the giant cell, it is quite possible that it relates to the endoderm.

What happens next is the enclosure of the big locomotive cell by a peculiar cap, which leads to the creation of an egg-shaped larva (Dr. 10); the discovery of this larva was crucial to Uliyanin's argument on *Amphigastrula invaginata*. Recently, however, his opinion has been considered completely erroneous; he thought that the rounded larva ended in a blastopore (mouth), but actually [the opening] is more like a narrow fissure extending along the entire body (Dr. 10, s). It is the result of closing the two ends of the cap, and its purpose is not to allow feeding (as Uliyanin suggested), but to remove pseudopodia from the protoplasm of the giant cell within the larva. This is why I do not share Uliyanin's view that the pseudopodia are offshoots of ectoderm cells. The larvae can swim well after withdrawing these organs inside the cell, which was not possible in the previous stage, which I have just described.

Rapid growth of the larva leads to changes in shape: it acquires a peculiar triangular form (Drs. 11 and 12). The fissure, which was previously straight, now curves and even branches (especially when the larva becomes triangular) (Drs. 11 and 12, s); this is linked to the need to produce pseudopodia. In this case there is no reason to consider this braching fissure as a homologue of the blastopore. It therefore follows that our larva has no gastrulation cavity; it is instead filled with the gigantic locomotive cell.

Transverse sections of the larval body (Dr. 13) show that it consists of two major layers and the gigantic inner cell. The ectoderm is clearly divided from the endoderm and consists of a single layer of narrow contractile cells (which are usually destroyed in preparation), The distribution of nuclei can be misleading, suggesting that the ectoderm is in two layers (*cf.* ectoderm in Dr. 14). The endoderm cells, previously aggregated, are now spread out in a single layer of epithelial cells, and are either flat or cylindrical.

Differentiation is already evident in these early stages. Thus, for instance, the longitudinal section (Dr. 13) clearly shows considerable thining of both embryonic layers in the region where the mouth of the first medusa will later be formed. A convexity (Dr. 14 and 15), which is the first rudiment of the proboscis of the medusa, is gradually developed by both embryonic layers; only the inner one, however, encloses the newly-formed cavity (which is the primary cavity of our animal). Its base consists of several cells which form a solid layer (Dr. 14, e), thus dividing the proboscis cavity from the upper end of the giant cell. This appears to be the only point of comparison between my data and Uliyanin's on differentiation of endoderm.

My observations confirm that division of merely a few cells, not the entire endoderm, is sufficient to create the above-mentioned partition. Uliyanin's drawings (5 and 10 of his Fig. 1) suggest to me that artificial damage has occurred as a consequence of inappropriate technique in the preparation of the larvae. This is quite obvious in his Dr. 5, which shows a chaotic assemblage of endoderm nuclei, with some of them completely uncovered – a phenomenon that I have never encountered before. The worthlessness of these sections is confirmed by their poor preservation.

I shall not concentrate on the already familiar phenomenon of differentiatiation but on the development of the giant cell. This cell exists for a long time after the developing animal attaches to the mouth of *Carmarina* (see Dr. 16). It lies in the dilated basal part of the animal, with the protoplasm and large nucleus clearly visible. The former contains several vacuoles crammed with small bodies resembling urinary corpuscles.[4] Perhaps this should be regarded as an early indication of degeneration which later leads to the destruction of the entire cell. This organ, i.e. the organ of movement, apparently becomes unnecessary after the immobilzation of our animal, and it therefore atrophies.

Regardless of the reason, the result is always negative: I have never encountered this cell in a completely developed adult animal, which buds prolifically. These observations, together with others that I have made, testify against particiption of the giant cell in the catching of food. [This] organism can feed only by means of diffusion of its own fluids, i.e. its body fluids; it can never grasp any solid particle. Overall, the evidence is against the presupposition of commensalism. Our budding animal is therefore a real parasite, and we have every reason to call it *Cunoctacantha parasitica.*

The entire life-cycle of this parasitic medusa appears to be a chain of secondary adaptations to its parasitic way of life. Perhaps the generational changes of *Cunoctacantha* are also secondary, since they do not really depend on metagenesis of hydroids. Asexual reproduction in this case is characterized by degeneration of the sexual organs, together with many other organs, which relates to young *Cunina rhododactyla* produced by budding. There is a great similarity between the asexual reproduction of the latter species and that of *Cunoctacatha*, as I mentioned earlier in 1872 [15].

In both cases we have sexually immature individuals that do not have edging bodies,[5] and they produce buds. While budding in *Cunina rhododactyla* is not significant and does not preclude the budded individual from futher development, it is intensified in parasitic *Cunoctacantha* to the extent that asexual reproduction hampers growth. Thus, generational changes in *Cunoctacantha parasitica* can be directly related to the method of reproduction in *Cunina rhododactyla*. This can also be shown by peculiarities in the formation of the larva in the parasite which are also secondary. It is thus difficult to relate the free-swimming, bilaterally symmetrical larva of *Cunoctacantha* (Dr. 10) to any primary larval form, especially as radial or biradial design is so typical of coelenterates.[6] As there is no gastrulation cavity, it cannot even be considered as a gastrula. It is more comparable to a form which I described earlier, the amphiblastula (Dr. 7–9) (although it is difficult to determine its real significance without knowing where the endoderm cells come from, as well as the histological status of the giant cell). My studies merely permit me to conclude, firstly, that the archigastrula ascribed by Uliyanin to the parasitic *Cunoctacantha* does not exist in reality, and, secondly, that the development of this medusa demonstrates a number of secondary adaptations.

These facts serve as confirmation of my previously expressed opinion that the genuine Metazoa have lower forms that develop by omitting the gastrula stage [16]; this is closely connected to the view that a form of parenchyma was the earliest type of endoderm. The development of endoderm in *Liriope*, described by me, is of the utmost interest precisely because a parenchyma stage occurs. Balfour [17] shares my opinion, accepting delamination in the lower Metazoa, and also considering it possible that the alimentary canal and mouth were of later phylogentic origin.

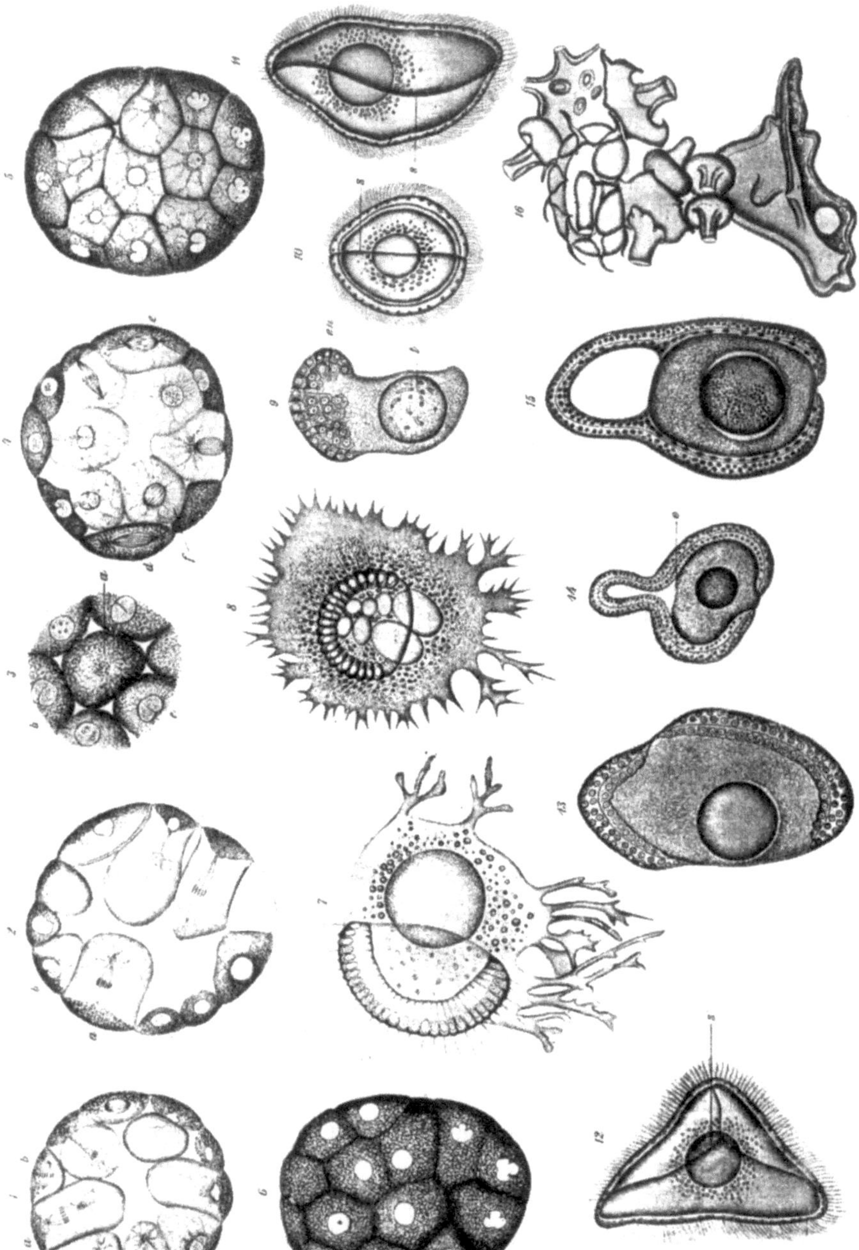

Figure XII

FIGURE XII

Dr. 1. Embryo of *Carmarina fungiformis* at the beginning of the formation of endoderm. Preparation is treated with the acetic acid. Magnification: 3+5, Gartnak's ocular.

Dr. 2. Embryo of *Liriope* at the beginning of delamination. Preparation is treated with the acetic acid. Magnification: 3+7; lower cell is not depicted in details because of a foreign body glued to it from above.

Dr. 3. Several cells of the same embryo from the upper layer. Letters *a, b, c* signify the same cells. Magnification: 3+7.

Dr. 4. The following stage of the formation of endoderm in the same species of medusa. *E, d* – cells dividing radially; *f* – an endodermic cell at the very beginning of its division. Preparation is treated with the acetic acid. Magnification: 3+7.

Dr. 5. Stage of parenchimella in *Liriope*. Optical section. Preparation is treated with the acetic acid. Magnification: 3+7.

Dr. 6. The same embryo from the upper layer.

Dr. 7. The earliest from the observed by me stages of *Cunoctantha parasitica*. Magnification: 3+7. A live animal is depicted.

Dr. 8. The same stage of the gigantic cell with two nuclei. Larva is killed by the osmic acid and then colored by picrocarmin. Magnification: 3+5.

Dr. 9. Section of the same stage. *En* – endoderm, *t* – pipe-liked figures inside the nucleus. Magnification: 3+5.

Dr. 10,11,12. Three larvae of *C. parasitica* on the following stages of developing. Live animals are depicted. Magnification: 3+4. *S* – a fissure that serves as a channel for pseudopodia to come out.

Dr. 13. A longitudinal section of the same larva. Magnification: 3+4.

Dr. 14. Cross-section of the more developed larva already in a stage of proliferation. *E* – cells of endoderm already separated. Magnification: 3+4.

Dr. 15. Cross-section of the same larva. Magnification: 3+4.

Dr. 16. Rudimentary sexless medusa already supplied with many buds. Gigantic cell is depicted in its base. Magnification: 3+4.

These considerations (together with those from my work, quoted above) called by some authors "the parenchymella theory", agree with my other observations on the life-history of the Metazoa. I shall continue to explore them in my further discussion.

Odessa, 7 (19) September 1881

COMPARATIVE EMBRYOLOGICAL STUDIES III.
(WITH FIGURES XIII AND XIV)
III. ON THE GASTRULA OF SOME MULTICELLULAR ANIMALS

Although early comparative embryology based its generalizations largely on the theory of embryonic layers in vertebrates, it was forced to adopt an alternative way of thinking and to turn to the the initial stages of the formation of embryonic layers in the lowest multicellular animals. The lower Metazoa provided solid ground for the search for information on the primary tissues, and the alimentary canal, the most ancient and widespread organ, played an outstanding role in acquiring such knowledge.

In 1847, Derbe [18] described the formation of the larva in *Echinus esculentus* and mentioned, in passing, a stage with a double layer of epithelium with a caecum opening to the exterior. This form of larva, depicted by Derbe in his Dr. 14, was named "gastrula" 25 years later by Haeckel, and this soon became accepted by the entire scientific community. According to Derbe, the opening of the enteron of the embryo was the mouth; in 1849, August Kröne [19] showed that this opening corresponds to the anus of the pluteus larva. This researcher was also the first to describe the peculiar method of formation of the embryonic gut, later called embolic invagination.

"This process looks exactly as if this embryonc enteron is formed as a result of [part of the] dermis gradually sinking into the body cavity and becoming concave. This depression becomes deeper and deeper, and gradually elongates into a canal which finally reaches that part of the wall of the body cavity facing its anterior surface. The blind sac thus created, which extends throughout the entire body cavity, is nothing less than the embryonic gut; the site of primary ingrowth becomes an aperture, which is in fact the anus, leading to the canal."

Bush [20] found gastrulas in *Asteracanthion glaciale* and *Echinocidaris neapolitanus,* but mistakenly took the concavity for the mouth. In the same year, Kröne [21] described a young auricularia[7] with the embryonic enteron in the form of a tube ending in an anus; he drew attention to the great similarity between this stage and young larvae of a sea-urchin. A gastrula was thus demonstrated for three major types of Echinodermata. Several years later, Kröne [22] saw exactly the same sort of larva in *Pelagia noctiluca.*[8] In this medusa, he found two-layered embryos, each with a slightly restricted stomach and a mouth aperture. He described only these stages, not taking into consideration their similarity to larval forms found by him in the Echinodermata.

Gastrulas were also discovered in some other representatives of the Metazoa. As early as 1853, Cobbold [23] sescribed a sac-like embryo of *Actinia mesembryanthemum*, but he did not pay sufficient attention to the layers constituting this embryo. At the beginning of the 60s, Alexander Agassiz studied a gastrula stage in *Asteracanthion beryllis*; the same stage was later discovered by Lerbullie in such highly developed forms as pulmonate gastropods.

These examples, however, seemed to be isolated facts, which also contradicted the widely accepted theory on the formation of embryonc layers in vertebrates. That is why these facts were of no significance in comparative embryology until Kovalevsky succeeded in showing that many lower vertebrates and invertebrates also have gastrulas. Kovalevsky found [gastrulas] not only in *Amphioxus* but also in *Cerianthus*, *Lineus*, *Phoronis*, *Ascidia*, *Aurelia* and a rain worm, so that already in 1866 he presumed that "this way of development (*i.e.* formation of enteron by invagination) is quite common and perhaps inherent in all animal's with a cleavage cavity." [24] He then confirmed and developed this opinion. [25]

Formation of both layers (external and internal) occurs in exactly the same manner in all the larvas mentioned here (*Amphioxus*, *Phoronis*, *Lineus*, *Asteracanthion*, *Ophiura*, *Echinus*, *Sagitta*, *Ascidia*, *Escholtzia*, *Cestum*, *Sepiola)*. The cleavage cavity and the layer of cells surrounding it, *i.e.* the blastoderm, which becomes concave and is the embryonic alimentary canal, form similarly and follow the division of the egg in all these larvae. They all have a cleavage cavity and nutritive yolk which fills it and then transforms into the [body] cavity of the future animal. As for the aperture formed by invagination, it becomes the anus, while the mouth aperture is formed by fusion of the invaginated layer of cells and the body wall; this aperture is positioned exactly on the site of this fusion.
Correlation of all these facts suggests the recognition of some major plan, for some forms at least." [25]

In the same year, Kovalevsky defended his dissertation and claimed that "the alimentary canal in animals whose eggs divide completely is formed by invagination." Convinced of the general similarity of all gastrulas, Kovalevsky was positive that the aperture produced by invagination becomes the anus, not the mouth. In his doctoral dissertation, [26] he went so far in this direction as to claim that this aperture in *Phoronis* also becomes the anus, and that the anterior part of *Actinotrocha*,[9] which has a little cap, is the anal part of the larva.

Quite recently, when this idea was rejected when it was established that, in *Phoronis*, the invagination aperture becomes the larval mouth, Kovalevsky went to the other extreme and claimed that such a transition also happens in *Pentacta doliolum*[10]. [27] By this declaration, he admitted, in fact, that there is no great similarity between different gastrulas. He also established that in some animals, like *Lumbricus* for example, the invagination aperture, or blastopore as it has recently been called, becomes the inner aperture of the esophagus, while in other animals it becomes a definite anus. He corrected his previous data on the transformation of the blastopore into the anus in *Amphioxus*, *Sagitta* and *Ascidia,* and admitted that the blastopore disappears in these animals, so a definite anus seems to be a new formation.

In his further embryological studies, Kovalevsky made an important discovery, namely that this process is quite similar in *Euaxes*,[11] where endoderm is created quite differently, not through invagination. "We can compare the enclosion of digestive gland cells in *Euxaes* with the same process in the Lumbricinae and in *Sagitta*; the only difference is that the considerable development of the lower layer does not allow it to invaginate, so cells of the upper layer have to overgrow the lower cells. It is quite certain that we should regard both the surrounding of the large, digestve-gland cells in the lower layer in *Euaxes* and invagination in the Lumbricinae as different extremes or stages of the same process." [28]

Kovalevsky thus had to discard not only his idea on the same destiny of the blastopore in different animals but also his earlier opinion on the identical method of formation of endoderm in all animals with complete cleavage. As early as 1868 [29] he observed formation of endoderm cells by detachment from the blastoderm where there was definitely no invagination. More recent studies confirmed his suspicions regarding the universality of invagination.

It is quite clear from all the above that Kovalevsky was quite persistent in his attempts to subject the facts of endoderm formation to some general law, but the experimental data did not allow him to do so. After his unsuccessful attempts to find invagination in some lower coelenterates like the Siphonophora and the Geryonidae, he should have been convinced that even the lowest and most primitive types of animals are not regulated by any general law of endoderm formation. It is perhaps no accident that his interesting and informative *Observations on the Development of Coelenterates* [30] does not contain any general review, although he promised one at the beginning.

It seems that all the above-mentioned difficulties were not taken seriously by Haeckel, who, relying on Kovalevsky's studies together with his own quite superficial observations, found it possible to go ahead with his gastraea theory, which is now quite well known [31].

Haeckel accepted Kovalevsky's idea, stated in 1874 but never repeated in later works, in an even more definite form than was originally suggested: "Acceptance of this ancient form (gastraea) is fuly justified by its homology, or morphological identity, in most animals of different types." (p. 20). There was no attempt on Haeckel's part to prove this identity; he never seriously considered data on the very different fate of the blastopore in different animals. Discussing the significance of the gastraea theory, he claimed that "the mouth aperture in the gastrula seems to be inherited in zoophytes only, and perhaps in a small group of worms. It appears to be present also as the Rusconian[12] anal aperture in vertebrates." (p. 42). He does not mention, however, that, according to Kröne, Aggasiz, Genzen, Kovalevsky and myself, the blastopore changes into the anus in the Echinodermata, which should throw serious doubt upon the assumed, and widely accepted, "unifrom identity" of all gastrulas.

On this and many other aspects of his theory, Haeckel relies upon Kovalevsky's early texts, namely, those of 1866–67, and ignores his later additions and corrections.[13] For instance, he regards delamination as not sufficiently proved and certainly not an important phenomenon. Haeckel, however, borrowed from Kovalevsky's later works the above mentioned idea on modification of endoderm development in *Euaxes* triggered by reinforced growth of nutritive yolk, and expanded it into the archigastrula theory. Haeckel's discogastrula and perigastrula theories are authentic, but the embryonic forms are very heterogeneous, as was earlier partly demonstrated by Balfour. [32]

Although the whole gastraea theory is clearly below the embryological standards of the 70s, Haeckel's influence can still be regarded as helpful in attempting to popularize (and sometimes simplify) Kovalevsky's work, especially taking into account that embryology at that time did not have a complete and/or comprehensive theory. However, numerous studies of initial stages in development in different animals which were carried out at this time, especially in Germany, encountered difficulties ignored by Haeckel. This brought out the obvious deficiencies of his theory. The most important is certainly the question of the blastopore. Since there is no doubt that the blastopore plays different roles in different animals, the very idea of the homology of all gastrulas can no longer be accepted. An attempt to accommodate this idea within the new experimental framework was undertaken by Bütschli and Hatschek (1876).

Bütschli [33] speculated that "hypothetical gastraeal ancient animals, ancestors to all Metazoa, had an invagination aperture corresponding to both the mouth and anal apertures in the Metazoa not only physiologically but also morphologically; this means that such apertures initially appeared by differentiation of the simple aperture in the gastraea and developed through partial fusion of its borders.." He applied this hypothesis to the special case of the Gastropoda, where the blastopore is extremely multifunctional.

Hatschek [34] developed similar ideas, but in a more general form and applicable to all Bilateria. He assumed that the mouth of the gastrula closed along a special line, the gastrular slit, which corresponds to the mid-line of the neural side of the animal, developed later. This closure reverses the initial formation of the aperture, so that the blastopore is quite large at first and only a small part of the residuum gradually transforms into the mouth. In some cases the mouth of the gastrula disappears completely. According to this theory, the gastrula in the Bilteria reveals a bilateral design very early (perhaps from the very beginning), manfested explicitly in this slit-like form of blastopore. Thus, the importance of a plan in the early embryonic stages was emphasized for the first time; this benefitted comparative embryology enormously.

Hatschek repeated and developed his ideas in other publications. Having outlined [the concept of] bilateral constitutional design, he came to the conclusion [35] that this design "is apparent in in the ovule of all bilaterally symmetrical animals." Although he thought that "the Bilteria descended from a unlateral gastrula", he still considered it possible that "not only the gastrula but also the embryonic bubble [= blastula] and even the ovule in bilaterally symmetrical animals to be bilaterally symmetrical." He observed a slit-like blastopore in *Pedicellina*[14] [36] and Annelida [37].

In his recent work on *Amphioxus*, he mentioned facts that are not quite consistent with the above. Thus he found bilateral symmetry only "in the stage of completed invagination." [38]. He did not attempt to interpret the earlier stages (Drs. 22 and 23) as bilateral, and there may be some reservations even in the later stages. Bilateral symmetry in a developed[15] gastraea is visible "only if viewed from the side, and only if the curvature is uneven, and also if the embryo is viewed from the side of the gastraeal mouth; only then is the outline seen to be oval and not circular." (p. 30). Small irregulrities in curvature are quite common, while bilateral symmetry is not easily visible; a blastopore which is oval "to some degree" is not an indication of bilateral constutional design since it can be seen in most biradial animals, like the Ctenophora. It is important that Hatschek does not provide us with drawings of a mature embryo of *Amphioxus* viewed from the side of the gastrula. Such drawings would be very helpful in understanding his views on closure [of the blastopore] in gastrulas.

While some proponents of the gastraea theory, like Rable [39] for example, consider major problems of this theory to be still unsolved solved, others, like the Hertwig brothers [40], pretend that "it gives a complete explanation of the two-layered embryo." They considered this claim to be an exhaustive argument, but they gave no data on the shape and closure of the blastopore in the Chaetognatha; thus the gap left by Kovalevsky and Bütschli is still not closed. Among the Herwigs" drawings of *Sagitta*, there is not even one depicting the blastopore in its entirety.

This short summary appears sufficient to show that the morphology of such an important embryonic stage as the gastrula is far from being understood, yet it is still possible to appreciate the phylogenetic significance of this embryonic/larval form. Adopting this as my starting point, I undertook a series of observations primarily on early forms of gastrula, the so-called archigastrula. I studied the gastrulas of Echinodermata because they are, in some respects, typical. Among worms, I chose relatively primitive forms, such as *Lineus*, *Polygordius* and *Phoronis*, then I turned to simple ascidians, whose gastrulas are frequently studied. My investigations took place mostly in the winter of 1880 at the zoological center of Naples, and partly at Messina (*Lineus*), and Odessa (*Polygordius*).

It is now my intention to present the results of my studies; I shall start with *Echinus microtuberculatus*. In 1879, Zelenka [41] studied the development of these Echinidae (under a different name, *Echinus miliaris*), but he did not pay attention to the problem of particular interest to us now, namely the problem of the blastopore. The first differentiation of embryonic cells in *Echinus*, as in many other animals, is initiated by thickening of the lower blastoderm cells, in the form of elongated multisided prisms, which gradually transform into much shorter elements than the remaining blastoderm. Before invagination, these thickened cells, as is well known, give off a number of free cells, which are then used to create cutis [42] and identified mesoderm and mesenchyme cells.

Zelenka claimed that these cells originated as two-cell groups resulting from the division of thickened blastoderm cells, and that their initial symmetry is largely preserved in the later stages. My observations do not confirm this, because I never detected a symmetry in primary mesoderm. As soon as mesoderm cells become separate elements (Dr. 1), they form a conglomerate of cells of different shapes, not yet penetrating the cleavage cavity and showing no trace of any symmetrical pattern. Mature mesoderm cells are more numerous and are much easier to distinguish from the surrounding thickened blastoderm cells; gradually they form a separate group of cells inside the cleavage cavity (Drs. 2, 3).

By taking into account a number of such stages, one can find several individuals with paired mesodermal elements; these, however, are exceptions, as was shown by me from serial sections. Nevertheless, I could not decide from cross-sections how mesodermal elements are initially created, either from the thickened blastoderm cells or through cell migration. I certainly saw a funnel-shaped chink that lengthened in the thickened part of the blastoderm, but I was never able to link it to the assumed symmetry of embyonic mesoderm.

When I refer to the developing free cells as either mesoderm or mesenchyme, my reason is that I cannot share the opinion of the Hertwig brothers [43] that mesoderm and mesenchyme are "completely different". Without here criticising the coelom theory, I should mention only several reasons to justify my opinion. I consider mesoderm and mesenchyme to be inseparable in a number of embryonic forms, for example in many annelids, arthropods and vertebrates; they cannot in principle be viewed as something different.

Although sharing the completely reasonable view of the Hertwig brothers [44] that "in order to understand the processes underlying the structure of organisms, one has to start initially with revelations of primordial simplicity played in front of the observer," it is obvious that one has to observe first the simplest multicelled organisms, the sponges, which are completely ignored by the Hertwig brothers. Sponges, even in the adult form, reveal a transition between separate endoderm cells and free elements of mesenchyme, and then these latter [change] into plaited epithelium of the gastrovascular canals [45]. The linings of these canals, commonly regarded as "endoderm", are considered by the Hertwig brothers as being primary stages of genuine mesoderm. Mesenchyme cells in the Echinodermata are regarded by the Hertwig brothers as nothing other than detached parts of the embryo, including both ectoderm and endoderm. From an entirely histological viewpoint, "the drastic differences between the two forms of tissue, epihelium and mesoderm," claimed by the Hertwig brothers [46], are as difficult to prove in the analysis of primordial stages of mesoderm as, for instance, the difference between cartiliginous endoderm cells in the tentacles of hydroids which form a solid "pencil", and the epithelium of endoderm cells which line the tentacle canals in many medusae. Here I have to repeat an objection made by Balfour [47], namely that even embryonic endoderm is often regarded as being composed of cells analogous to mesenchyme which, however, do not reveal their epithelial nature.

Invagination starts approximately within 20 minutes of fertilization;[16] some of the thickened blastoderm cells in the lower part of the lava take part in it. Gastraeal invagination, initially insignificant and hardly visible from the surface, becomes more definite and then becomes a blastopore with more or less regular circular contours (Dr. 4) [48]. The diameter of the blastopore is almost one third that of the lower part of the larva; both these formations are circular. The size of the blastopore gradually diminishes in subsequent development; the blastopore itself, however, keeps its initial shape and its central or subcentral position. This shows that its margins grow absolutely evenly.

After the invaginated endoderm sac rises to approximately half [the height of] the jelly bag [blastocoel], the radial design of the larva starts changing to bilateral. This is demonstrated by a significant curving of the upper part of the larva (Dr. 9); this results in a distinction between the ventral surface, which later generates the mouth, and the more convex dorsal surface.

Bilateral symmetry becomes more apparent with development of the skeleton, which takes place, and I want to emphasize this, within the mesoderm cells (Drs. 9–11). A little later the endoderm sac bends to the upper part of the ventral surface (Dr. 12), finally effecting bilateral symmetry of this organ. The blastopore, by contrast, preserves a position corresponding to its radial design; its circular shape becomes even more apparent, in spite of a considerable reduction in size. Its size keeps diminishing during subsequent development, becoming only one third of its initial diameter (larva shown in Dr. 13). It transforms, as is widely known, into the anal aperture, so this is an example of a so-called anal gastrula. The mouth aperture (Dr. 13, a) is so far from the blastopore and its place of appearance appears so early that it makes no sense to compare it to the original, much bigger blastopore. Thus the theory of the slit-like blastopore cannot be applied to our animal, despite the fact that the larva acquires its bilateral symmetry so early.

Let us now consider worms. Hatschek [49] thought that he had found, in the Annelida, "a comparatively well preserved" form of the primitive gastrula; this became the basis of his theory, mentioned above. The life-history of earthworms, however, shows such amazing features of adaptational abbreviation that this view can hardly be accepted. Unequal cleavage and the absence of metamorphosis suggest secondary developmental phenomena This is quite aufficient reason to give preference to other Annelida, with a longer period of development. My major expectations, however, are with the Archiannelida, such as *Polygordius*. Eggs of these animals (in particular *P. flavocapitulis*, widespread in the Black Sea) undergo a period of regular division of yolk, which leads to a blastula and then to an invaginal gasrula. The larva hatches as early as the blastula stage, and continues its development to a specific form of trochophore [50]. After significant thickening of some blastoderm cells, the embryo invaginates; the invagination cavity, although initially still small, has quite distinct edges, so that the blastopore is easily visible. The blastopore, viewed facing the concavity in the blastula, is circular and median, and has a diameter of apprximately one third that of the embryo (Dr. 42). The similarity of such a gastrula to that of *Echinus*, described above, is quite remarkable (*cf.* Drs. 42, 43 with Drs. 4, 5). In both cases we have a gastrula constituted according to a radial design, with a three-layered rounded blastopore. The cleavage cavity of *Polygordius* contains some free cells, which quite obviously represent initial elements of mesoderm that quite possibly originated from endoderm (as was earlier stated by Repiahov).

Differences between gastrulas are all of a qualitative nature. The cleavage cavity in *Echinus*, the so-called jelly bag,[17] is considerably bigger, while the endoderm sac, by contrast, is considerably smaller than in *Polygordius*; all the ectoderm in the former is covered by vibrating cilia, but only part of it in the latter. Differences in shape between the two gastrulas is not very significant: in *Polygordius* it is ovoid and has a membrane which is visible because of a convexity in the region of the blastopore; this membrane looks like a watch-glass (Dr. 43 c). Subsequent development in this annelid gives the gastrula an elongated shape; the endoderm sac enlarges in a regular manner (Dr. 44), while the blastopore becomes smaller and smaller. The latter preserves its round shape or becomes more oval, in conformity with the general radial design of the larva. The next stage in the development of the blastopore does not change its shape; the blastopore merely moves deeper into the body of the larva which finally results in the formation of an oesophagus.

The latter organ is of endodermal origin and consists of a vast tunnel-shaped body with a round or oval mouth aperture, quite definite and comparatively large, which is the exterior opening of the oesophagus (Dr. 45 a,b). As a result of the curvature of the oesophagus, the larva loses its initial radial shape and acquires strongly marked bilateral symmetry. This design is apparent not only in the shape of the entire body but, even more so, in the enlargement of the endoderm sac into the mid-gut which is now close to the ectoderm. The mouth aperture, reduced by constriction of its margins, still preserves its shape in the lower half of the larval body. The larva can still swim with the incredible speed of an arrow. As for the alimentary canal, it remains blind for a time, but then acquires a short hind gut and an anal aperture.

It is remarkable how the initial similarity betwen the gastrulas of *Polygordius* and *Echinus* changes to total difference in their subsequent development. In the case of *Echinus* we have a typical radial anal gastrula, while in *Polygordius* we encouter a gastrula, still typical and radial, but oral. Now it is difficult to find sufficient justification to claim that the gastrula of *Polygordius*, defined in general terms, is a more primitive form than some gastrulas of the Lumbricinae or those of the Polychaeta, or that the radial shape of the gastrula of *Polygordius* is secondary, developed from an initial bilateral design. I feel confident in assuming that what is here under consideration is a certain primordial embryonic form from which later, through secondary adaptation, the characteristic features of the gastrulas of other Annelida have developed.

Incidentally, I think that the big slit-like blastopore is also a late acquisition of the gastrula; it seems to develop as a result of premature differentiation to a bilaterally symmetrical shape [induced] by a powerful endoderm. Such a blastopore has been described, for example, by Jiar [51] in *Salmacina dysteri*, by Repiahov [52] in *Saccocirrus*, and by Goetthe [53] in *Nereis*, and can also be seen in Oligichaeta. All these forms, however, cannot compete with *Polygordius*, at least as far as primordiality is concerned.

Schizogastrulas[18] were futher studied and described in Nematoda, Brachiopoda, Gastropoda, Crustacea, and Insecta, and also in animals placed phylogenetically higher than *Polygordius* and that do not undergo such an extended life-cycle as the above mentioned annelid.

Gastrulation in *Polygordius* is even more significant when considered in relation to some results of studies on lower Vermes.[19] If we ignore forms with irregular division (like, for example, a dendrocoel) to avoid too many secondary phenomena, and turn exclusively to unsegmented worms[20] with regular initial cleavage, blastula formation, and clear invaginal gastrula, the first choice of study is the Nemertea,[21] with pilidium larvae.

Because my previous studies on *Lineus facteus* [54] were not sufficiently complete to give an accurate account of the blastopore, I turned to this animal again. I have to add to what was earlier said about the eggs that each egg is surrounded by a jelly-like membrane. As a result of frequent adhesion between such eggs, they can form small aggregations which sink into deep waters. Complete and regular division of the egg leads to the creation of a blastula with a rather narrow cleavage cavity (Dr. 15).

Blastular cells, initially irregular, are later made up of two kinds of elements; cells in the lower half of the blastula (endoderm cells) are thicker and bigger than cells in the upper half (ectoderm cells). Several mesoderm cells, in the cleavage cavity next to endoderm cells (Drs. 16–18, m[22]), apparently (if not certainly) originated from ectoderm cells, a process very similar to that in the Echinodermata. The blastoderm cells become covered with short cilia quite early, so that the blastula can revolve within the egg membrane. The embryo then acquires a characteristic shape with a clearly radial design, so well illustrated in Dr. 17. At this stage, the lower flattened surface supports higher endoderm cells which soon start to invaginate.

At this stage, the larva often tears the egg membrane and starts to swim in the water as a free blastula. More often, however, it stays within the membrane for some time and hatches only when it turns into a gastrula. Invagination occurs evenly over the entire lower surface, and the quite massive blastopore (Dr. 19) acquires a more or less perfect circular shape. It is placed right in the center of this circle (or close to it).

The young blastula (Drs. 19–21) is initially radial (ignoring some slight irregularities in the contours of the blind endoderm sac). Such a radial gastrula is similar to the oral gastrula of *Polygordius*; the differences between them are insignificant. The radial design, however, does not last long; bilateral symmetry becomes more and more apparent with the subsequent development of the gastrula. It can be noticed first in lateral view [which shows] a curved endoderm sac (Dr. 24); the blastopore, which becomes smaller at that time, changes from rounded to egg-shaped, with both narrow and wide poles clearly visible (Dr. 22). With protrusion of the oesophagus, the endoderm sac becomes the mid gut and shifts progressively so that it finally rests against the rear wall of the pilidium (Drs. 25, 26). As a result of the formation of the oesophagus, the blastopore is pulled inside, while the mouth aperture appears in its place and gradually grows (Dr. 27). We now see the characteristic features of a pilidium, partially known from previous studies.

In comparing the pilidium gastrula with the above-described developmental stages of *Polygordius*, we have to emphasize firstly their amazing similarity. The pilidium gastrula is also initially radial, with a three-layered mouth, and the transition to biradial symmetry is determined by displacement of the endoderm sac. It can be broadly claimed that the larva of *Polygordius* (Dr. 45) goes through a temporary transitional pilidium stage, characterized, among other features, by a blind, closed mid gut. I provisionally assign the status of pilidium to the embryo of *Pedicellina*, described by Hatschek (pp. 16–20), and the embryo of *Teredo* (p. 15).

It follows, therefore, that the pilidium is a well known larva. Not sharing Hatschek's opinion [55] (namely that the pilidium is a sort of trochophore), I agree with Balfour [56], who considers the trochophore to be a descendant of the pilidium, a primitive but phylogenetically important type of larva.

The development of the gastrula in *Phoronis* is similar, in some respects, to that seen in the pilidium. Cells of half the blastoderm thicken to create the first embryonic endoderm. The initially vast cleavage cavity is reduced in size to a small chink (Dr. 29), and soon after that, endoderm formation begins. Mesoderm[22] in this animal, as in others discussed here, arises as a development of several endoderm cells (Drs. 30, 31, m). Invagination begins a little later and is initially so slight that it cannot be detected from the surface of the gastrula. Invagination is visible only when the oval embryo is viewed in optical longitudinal section.

A closer look reveals bilateral symmetry in larval structure already at this stage. Remnants of the cleavage cavity are bigger on one side than the other; that is because mesoderm cells are more concentrated on the former [side] (Dr 31, m). This becomes more and more obvious with subsequent deepening of the endoderm sac (see Drs. 30, 31). The blastopore, however, takes no part in this change to bilateral symmetry at this stage and preserves its rounded contours for some time (Dr. 32). Soon after that, it acquires the shape of a tapering oval (Dr. 34). Viewed from the right angle, the embryo shows a longitudinal fissure, starting from the blastopore and extending to the rear of the embryo, limited only by endoderm. This fissure makes the bilateral design of the embryo more apparent, but it is a transient structure, not visible in the later stages. The anterior part of the embryo is thus a zone of very intensive development which includes the body cavity, developed from the original cleavage cavity (Drs. 37, 38).

The size of the blastopore reduces quite significantly, while it becomes more heart-shaped; its outline, however, remains distinct only in the rear part and almost completely disappears in the front segment. This fact is linked to the hollowing out of the oesophagus, which takes place predominantly in the front part of the body (see Drs. 37 and 39). The above-mentioned fissure has already disappeared at this stage, but a new aperture, the anus, develops next to the lower pole of the body, on the ventral surface of the embryo (Dr. 38, an). There are already many mesodermal elements in the body cavity; these are found on both embryonic layers and later become a double layer of mesodermal epithelium (Drs. 37–39, m, m'). Further development leads to the formation of the previously known umbrella-like adjunct to the anterior end, which eventually completely covers the blastopore (Drs. 40, 41). At this stage, the embryo hatches as a larva and swims at large in the water. Cilia, indispensible for this, were previously developed at the gastrula stage.

Gastrulation in *Phoronis* shows that we have here a more primitive method of achieving bilaterally symmetrical design than in *Echinus*, *Polygordius,* or *Lineus*. We first encounter this type of design in *Phoronis* quite early [in development], at the time of invagination of the endoderm; there is, therefore, no point in referring to a radial gastrula after this time. Remarkably enough, the major role in this process is played by the endoderm sac, while the blastopore shows little indication of this [bilateral] design for some time.

The gastrula of *Phoronis* should be listed among the three-layered oral gastrulas, and considered a close relative of the pilidium gastrula. But even the youngest actinotroch is amazingly similar to the larva of Planaria (*e.g. Stilochopsis*), which is a form of pilidium, so it would be quite reasonable to regard them as more developed larvae of the same type, already with an anal aperture. Their similarity to larvae of the Echinodermata, noted by Balfour [57], should be regarded as a consequence of secondary adaptation; gastrulation here [*i.e.* oral gastrulation] is completely different from that in the Echinodermata.

The gastrula of *Phoronis* differs from other oral gastrulas principally in its bilateral symmetry; this difference must be linked to the abbreviation of development in this animal form, which is correlated with late and rapid metamorphosis. To test this proposition, first and foremost one has to study a so-called archigastrula, using as an example an animal with a shortened developmental cycle. This is the case in *Ascidia*, for it does not have a ciliated larva. The larva hatches from the egg only when it is relatively well developed, so there is every reason to regard this as a case of shortened embyonic development. I selected *Ascidia mentula* from this group because of the transparency of the egg and other important features. Although *Weitere Studien uber die Entwicklung der einfachen Ascidien* by Kovalevsky [58] is complete in many respects, it does not contain a comprehensive answer to questions on changes in form and the shape of the blastopore in different stages; so I have to present my own observations.

The last stage in which any radial design can be seen is a blastula with thickened vegetal region, shown in Dr. 46; it corresponds to the embryo of *Pilidium* shown in Dr. 17, because they both have mesodermal cells and cilia. Bilateral symmetry is already visible in this blastula before the formation of a gastrula is completed, and shows itself in a distinctive arangement of endoderm cells (Dr. 47). On each side one can see a pair of larger cells coming together. Bilateral design is also reflected in the irregular curvature of the entire embryo (Dr. 48), so already at this stage the head end of the embryo is clearly distinguishable from the tail region, in both anterior and posterior view. The next stage is characterized by very slight invagination, and bilateral symmetry becomes even more pronounced. The head region shows increasing differences from the tail region; even the rough outline of invaginated endoderm shows comparable changes (Dr. 50). Bilateral design is now apparent not only in the four above-mentioned cells but in almost all elements of both layers (Dr. 49). The entire embryo is divided into two symmetrical halves by a line composed of the edges of separate cells; these cells form an amazingly regular row on each side of this line.

The blastopore, distinguishable only from this stage on, occupies a major part of the surface of the gastrula and immediately reveals its bilateral design. It is almost horseshoe-shaped, while from above, *i.e.* in the head area, it is rounded. It is surrounded on both sides by the four above-mentioned large cells, and gradually flattens in the lower part, where its outline becomes almost indistinguishable (Dr. 49). Such a blastopore, which can by no means be considered chink-like, is quite similar to the blastopore of some molluscs and brachiopods, especially that of *Agriope* (see Kovalevsky).

While subsequent development of the notochord of the embryo is by continuing invagination, a transformation of the blastopore occurs; its lateral margins converge so that it becomes heart-shaped, while its size diminishes significantly (Dr. 52). One can observe this process under the microscope in a live animal, and draw its contours by means of a prism.[23] The two outlines reproduced in Dr. 52 were obtained in this way; they were taken over a period of an hour and a half. Comparing these outlines, it is easy to see that the upper and lower margins of the blastopore grow faster than the sides; this is also clarified in further development. The blastopore maintains its central position for some time, but with every subsequent step in development it is dragged deeper and deeper. *Ascidia mentula*, like *Phoronis*, has an ectodermal fissure starting at the blastopre and extending to the lower end of the embryo (Drs 56, 58, s), and later connected to the neural fissure (Dr. 60, s). In general, the described embryonic processes look quite similar in both animals, at least superficially; further development, however, reveals drastic differences.

The endoderm sac, for example, in *Phoronis* is pulled into the rear part of the body, while, by contrast, in *Ascidia* it enters the cephalic part; the blastopore in the latter animal is wholly in the lower part of the body, but in *Phoronis* it is in the cephalic part and passes into the inlet of the oesophagus. The blastopore in *Ascidia* diminishes in size, loses its previous shape and changes to an almost completely circular hole (Drs. 57, 59), which remains for some time the central outlet in the ring-shaped notochord (Dr. 57, ch). The growing edges of the neural fissure soon completely shield it from the observer.

Considering the gastrulas described above, that of *Ascidia* is the most relevant to Hatschek's theory, since here we encounter the earliest known development of bilateral symmetry and the largest and most symmetrical blastopore. Comparing the gastrula of *Amphioxus,* described by Hatschek, with that of *Ascidia*, it may be argued that the properties of the former (described above) should be considered to be a result of shortened development. No matter how hard Hatschek tried to find the earliest [radially] symmetrical design in the gastrula of *Amphioxus*, he could show its existence only in the second gastrula stage; even then, however, the blastopore remains oval, *i.e.* bilateral, for quite some time.

The first conclusion I have to draw now is that the complete correspondence of archigastrulas, proclaimed by Haeckel [59], does not exist in reality. The above-mentioned facts are sufficient proof of this. Further, I have to emphasize that the idea that a gastrula must necessarily be two-layered should also be rejected. We encountered quite primitive gastrulas which already have several mesoderm cells in the blastula stage. Two-layered archigastrulas are, in general, quite rare except in animals in which mesoderm originates in folds of ectoderm (like *Amphioxus* and *Sagitta*). Such gastrulas occur in metazoans like Acalepha[24] and *Ascidia*, where mesoderm appears fairly late, if at all.

It has already been emphasized that archigastrulas have a radial design not only in the Echinodermata but also in lower bilaterally symmetrical animals. Bilateral symmetry shows itself first in the general shape of the body, and only later does it influence the shape of the blastopore. In cases where the blastopore is bilaterally symmetrical from the very beginning, this shape is already apparent in the blastula; so it can be postulated that bilateral symmetry of the blastopore is merely secondary. Viewed from this perspective, a cleft-like large blastopore does not have any palingenetic[25] significance and should be regarded as a product of embryonic adaptation.

If a radial gastrula is regarded as primordial, the question immediately arises, are the superficially similar gastrulas of Echinodermata and lower Annelida[26] (*Lineus, Polygordius*) equivalent or not? A positive answer to this question unavoidably implies that the anal aperture in the Echinodermata is a homologue of the oral aperture and oesophagus of worms, which is risky, to say the least [60]. Such a daring suggestion could only be considered if the gastraea theory seemed adequate to solve major morphological problems. This, however, is by no means the case.

I have already emphasized several times that the invaginal gastrula theory encounters difficulties in the analysis of embryological observations on the lowest Metazoa. A peculiar method of endoderm formation, characteristic of the most primitive sponges (*Halisarca* and *Ascetta*) and hydromedusae (*Eucope*, *Obelia*, *Tiara*, *Zygodactyla* and others), is the ingression of some cells into the cleavage cavity, which leads firstly to the emergence of a solid parenchyma and only secondarily to the formation of an epithelial layer of endoderm. This method of creation of a parenchymella, through migration, cannot be regarded as shortening the process of invagination. Such migration of cells takes more time than invagination, and, in addition, it should certainly be regarded as more complete. By contrast, endoderm formation in the higher Metazoa (Acalepha, *Actinia*, some pectinates[27]) can unequivocally be regarded as abbreviated. In this latter case [61], not only is the endoderm sac formed instantly but the entire parenchymella stage is omitted, while the gastrulation cavity is produced through invagination.

Since we have no reason to consider the formation of endoderm through [cell] migration as a secondary phenomenon, we should account for it in any attempt to explain the ancestral embryonic process. Genuine delamination, seen in the majority of hydroids, siphonophores and many sponges, can best be interpreted as form of shortening of the initial migration process, which makes it even more typical than invagination.

When Balfour [62] claimed that "in certain cases at least, delaminating occurs secondarily from invagination", one should not forget that he was referring not to sponges or Cnidaria but to *Theridium* and *Nemertina*.[28] Further, Kovalevsky himself [63] confessed that he did not trace the origin of endoderm cells in *Theridium*; a glance at his drawings (Taf. III, Drs. 6–8) is sufficient to show that the origin of endoderm in *Theridium* was far from clear to him. This is even more the case in respect to his data on delamination in *Nemertina* (*Tetrastemma*), resulting, as it seems, from insufficiently accurate observation; this is quite understandable if we take into consideration that the direct development of *Nemertina* is one of the most difficult chapters in embryology. I dare to suggest that a re-examination of endoderm formation in *Theridium* and *Tetrastemma* (which is particularly important for comparative embryology) a gastrula stage will be found. The claim that delamination takes place in *Ophiothrix*, made quite recently by Apostolides [64], was based, it seems, on confusion between mesoderm and endoderm.

In order to believe in the gastraea theory, one [firstly] has to accept the morphological correspondence of the head of a worm and the rear end of a larval echinoderm, and [secondly] that the formation of endoderm in the lower Metazoa, and the parenchyma so often observed there, are inexplicable phenomena, without any phylogenetic significance. According to this theory, the lowest worms, *i.e.* the parenchymatic Turbellaria (the so-called Acoela), are degenerate descendants of some animal with a vast cavity [surrounding] the gut, although neither parasitism, nor a sedentery way of life, nor any other cause of degeneration are characteristic of them.

All these difficulties can easily be avoided if we admit that the gastraea is not the primary form of metazoan, but, on the contrary, it is a later form, following those without an alimentary canal and without internal digestive parenchyma (mesoendoderm). Further to this point, the parenchymal larvae of sponges and hydroids and of lower Turbellaria without alimentary canal should be regarded as close relatives both of each other and of the more ancient Metazoa, from a phylogenetic viewpoint. Only later did higher forms of animals, with a differentiated alimentary canal, appear, like, for example, contemporary hydroid polyps, the ontogenesis of which repeats major phylogenetic phases – migration of endoderm cells, formation of a solid parenchymella, and later development of an intestinal or mouth aperture.

The most primitive hydroids, like *Protohydra* [65], show the closest corresponce to the hypothetical gastraea, although their organization is not lower or more primitive than sponges or Acoela. As a result of shortened development, major features of these hydroids appeared earlier, which led to the gastrula as an ontogenetic stage. The existence of such a stage was sufficiently demonstrated by Ray Lankester [66]. The tendency to the possible early differentiation of fairly well developed organs (*e.g.* early formation of eyes in cephalopods, or the chorda in vertebrates) can be explained by the early and rapid development of endoderm in the Metazoa, which has enormous influence on the process of gastrulation. Thus we see how a gastrula with a gigantic blastopore could have developed (as in *Ascidia*, molluscs and insects), which, as already pointed out, is easy to interpret as a product of embryonic adaptation. This can also help to explain the occurrence of several gastrula stages in the development of the same form of animal.

Recently an entire "*embarras de richesse*" of gastrulas has come to notice. I refer here not only to vertebrates (like, for example, bony fish, where completely different methods of gastrulation were described by Kupffer and other embryologists), or some Arthropoda (where Balfour found secondary gastrulas), but to even lower Metazoa. The Acalefa also have two gastrula stages. The first may be observed when invagination of endoderm takes place, and thus is associated with the earliest embryonic stages; the second form of gastrula is created when the larva is setting; this is the stage of formation of the secondary mouth (ascula).

I cannot agree with Haeckel [67] when he considers the formation of a secondary mouth aperture in Discomedusae not as a real innovation but rather as another opening to a canal which has been closed for some time. We know that the gastrula of *Aurelia aurita* (the very same jellyfish studied by Haeckel) is separated from the ascula by a lengthy intermediate stage, in which the planula (as yet without a mouth) grows considerably and also develops stinging organs. In some cases, several individuals merge to create a colossal planula. The second mouth appears a long time after settling; this mouth, according to a correct remark by Haeckel [68], coincides with the location of the blastopore. I therefore see no reason to accept the existence of a mouth aperture throughout the entire planula stage, just as other animals that do have mouths behave differently while free-swimming.

To support my opinion, I can cite an analogous case from the field of sponges. *Sycandra* is of interest here because it undergoes three forms of gastrulation in its developmental cycle: first a pseudogastrula, described by F.E. Schultze, then an

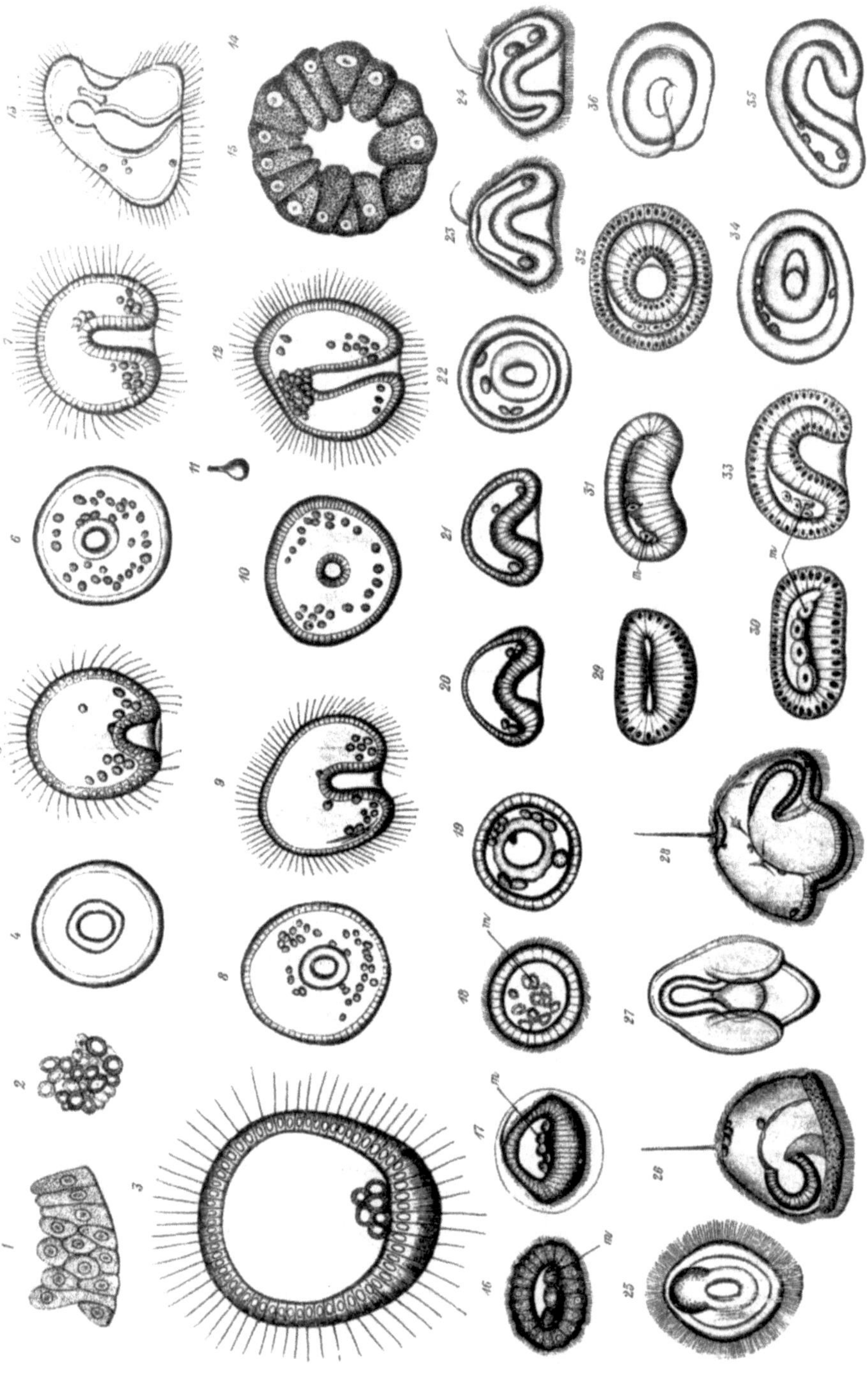

Figure XIII

FIGURES XIII-XIV

Drs.1–14 relate to *Echinus microtuberculatus*.
Dr. 1. Part of a lower end of embryo under-going formation of mesodermic cells. Ocular 3+. Syst. 9 (dry) Gartnak. Osmic fumes.
Dr. 2. A number of mesodermic cells of 18-hour old larva depicted from above. 3+7. Osmic preparation.
Dr. 4. 20-hour old larva in a stage of gastrulation depicted from above. 3+5. A live object is depicted.
Dr. 5. Same gastrula in optic longitudinal section. 3+5. A live object is depicted.
Dr. 7. The same object in optic longitudinal section. 3+5. A live object is depicted.
Drs. 8–9. More mature gastrula under-going formation of skeleton. 3+5.
Dr. 10. 23-hour old gastrula depicted from above. 3+5. A live object is depicted.
Dr. 11. Isolated skeleton-forming cell of the same larva. 3+8. A live object is depicted.
Dr. 13. 39-hour old pluteus under-going formation of mouth. Skeleton is depicted. 3+5. A live object is depicted. 0 – mouth deepening.
Drs. 15–20 relate to *Lineus lacteus*.
Dr. 15. Blastula. 3+7. Osmic acid. Glycerin.
Dr. 16. Following stage under-going mesodermic formation. 3+5. Acetic acid. Glycerin.
Dr. 17. Blastula with thickened endoderm. 3+5. Osmic acid. Glycerin.
Dr. 18. Same larva from above.
Dr. 19. Early stage of gastrula depicted from the surface. 3+5. Osmic acid. Glycerin.
Dr. 20. Same larva in optic longitudinal section.
Dr. 21. Same larva turned 90 degrees.
Dr. 22. Gastrula with two-axis symmetry depicted from the surface. 3+5. Osmic acid. Glycerin.
Dr. 23. Same gastrula. Frontal section.
Dr. 24. Same larva. Saggital section.
Dr. 25. 2-day old pilidium depicted from gastric surface. 3+5. A live object is depicted.
Dr. 26. Same pilidium. Saggital section.
Dr. 27.More mature pilidium depicted from gastric surface. 3+5. A live object is depicted.
Drs. 29–40 relate to *Phoronis hippocrepina*. All preparations dyed by acid carmin and placed in glycerin. Celia are not depicted anywhere.
Dr. 29. Blastula with thickened endoderm.
Dr. Further development of blastula with mesodermic cells.
Dr. 31. The earliest stage of larva.
Dr. 37. Following stage with the already disappeared furrow. M- visceric leaf of mesoderm; m' – dermic leaf.
Dr. 38. Same larva turned 90° degrees to its longitudinal axis. *An* – anus.
Dr. 39. Same larva in profile. *M*, *m'* – same as in Dr. 37.
Dr. 40. Larva with a large head palette depicted from gastric surface.
Dr. 41. More mature larva.
Drs. 42–45 relate to *Polygordius flavocapitatus*. Preparations are treated by osmic fumes.
Dr. 42. Gastrula depicted from the gastric surface.
Dr. 43. Same gastrula in optic longitudinal section. 3+9. C – swelling of egg.
Dr. 44. More mature gastrula with initiated swelling of gullet. 3+9.
Dr. 45. Larva with swelled gullet. 3+8.
Drs. 46–60 relate to *Ascidia mentula*. All depicted larvae are alive, 3+9 (with the exception of Dr. 53).
Dr. 46. 32-stage blastula.
Dr. 47. Last stage of blastula with a symmetrical design of cells. Depicted from gastric surface.
Dr. 48. The same stage in optic longitudinal section.
Dr. 49. Early stage of gastrula depicted from gastric surface.
Dr. 50. The same in optic longitudinal section.
Dr. 52. Same larva ¼ hours later.

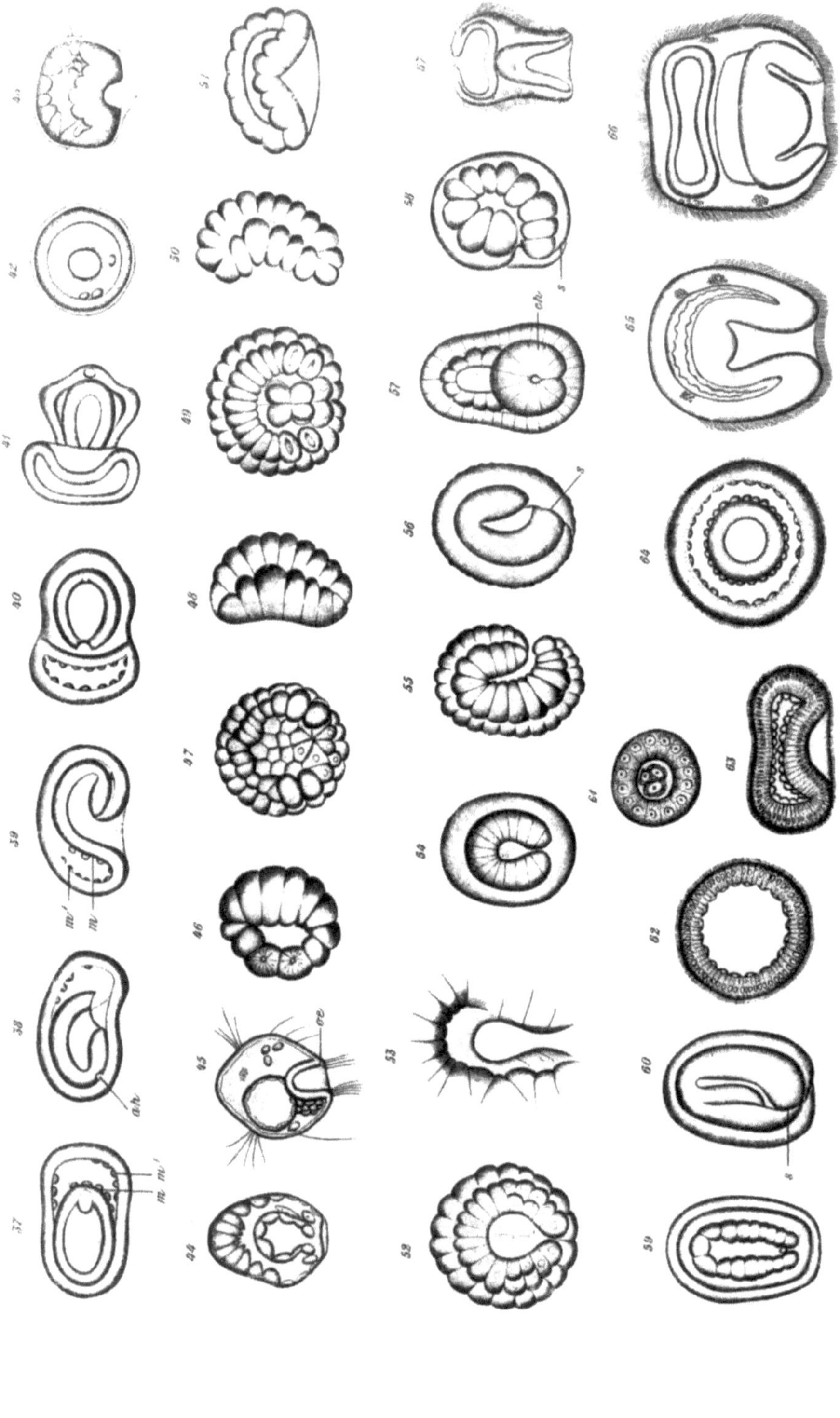

Figure XIV

Dr. 56. Same larva in a position chosen to show furrow s.
Dr. 58. Last stage of gastrula with a small blastopore and two rows of nervous cells.
Dr. 60. Same larva with furrow s.
Drs. 61–67 relate to *Discoporella radiata*.
Dr. 63. Pseudoblastula in optic section. 3+8. Acidic acid.
Dr. 64. Following stage depicted from a surface of pseudogastrula. 3+8.
Dr. 65. More mature stage of pseudogastrula with a deep swelling of a sucker.
Dr. 66. Larva with a developed sucker. 3+7.
Dr. 67. Fastened larva with a mouth opening and a partially swelled sucker.

amphigastrula, and, finally, the youngest form of sponge with a mouth. (The presence of mesoderm cells cannot be regarded as an objection because such cells are encountered in other gastrulas.) There is yet another peculiar stage without a mouth inserted between the two later stages; the osculum bursts open long after settlement, *i.e.* after the formation of the skeleton. This example is also evidence that abbreviation in development, presupposed above, was not instant but gradual. The initial stage of abbreviation extended to the premature formation of endoderm and its invagination into the larvae. Under such circumstances, it becomes self-explanatory that, of several gastrulas, at least one fulfills a function that is a more or less wholly adaptive. Discrepancies between various gastrulas, so difficult to explain in terms of the gastraea theory, can, under my theory, be regarded as examples of different embryonic variations in such close relatives as Isopoda and Amphipoda.

If some animals undergo hypergastrulation, others can provide examples of a pseudogastrula, *i.e.*, a stage [superficially] similar to a gastrula but not corresponding to it at all. Koelliker [69], for example, claimed that the gastrula described by E. van Beneden in the rabbit does not correspond to this stage because the final ectoderm develops from van Beneden's endoderm, etc. I spoke against regarding the stage described by van Beneden in *Dicyema* [70] as a gastrula; in this case we are quite possibly dealing with a pseudogastrula. An interesting example of this stage is provided by the *Bryozoa cyclostoma*. Stages described as gastrulas in several species (*Phalangella*, *Crisia*, *Diastopora*) are also clearly pseudogastrulas; I had an opportunity to verify this when I studied various Cyclostomata, *Disporella radiata* in particular.

Endoderm is formed much earlier than was assumed by Barrois [71]; I was able to detect it, as yet underdeveloped, even in the earliest embryonic stages observed by me (Dr. 61). Endoderm cells within the embryo form a solid layer of thickened cells (Dr. 62), so that this stage represents the so-called diblastula[29] (*i.e.*, a stage corresponding to genuine gastrulation, which normally should have occurred long ago). Half the diblastula now invaginates in a similar manner to that occurring in the formation of a radial invaginal gastrula (Dr. 63). The invagination cavity is initialy very shallow, while the pseudoblastopore is very round (Dr. 64). Further invagination causes pressure on the endoderm, so that the intestinal cavity becomes a chink-like slit (Dr. 65). The described embryo at this stage can easily be mistaken for a three-layered gastrula, although a suspicious protrusion is already visible at the end of the invagination cavity.

Subsequent development soon shows that we are here dealing with the so-called sucker, a common formation in bryozoan larvae (Dr. 66). The invaginated sac, therefore, is an ectodemal structure, which is predominantly true of the origins of the sucker itself. When the larva settles, the sucker comes out of its covering and serves to anchor the larva. The genuine endoderm sac in the developed larva still remains closed for a while, but it bursts open as a central round hole. Thus a secondary form, similar to a gastrula and supplied with a sucker, finally appears (Dr. 67).

Odessa, 2 (14) May 1882.

COMPARATIVE EMBRYOLOGICAL STUDIES IV (WITH FIGURES XVIII, XIX AND XX) IV. ON THE FORMATION OF THE GASTRULA AND MESODERM IN CTENOPHORES[30]

Since gastrulas and mesoderm are not characteristic of all Metazoa but of the higher members only, knowledge of metazoan development is crucial to understanding the first stages of gastrulation and also the formation of mesoderm.

Despite universal acceptance of a gastrula stage in the development of ctenophores, the circumstances of its formation and development are very little known. Kovalevsky assumed that the gigantic blastomeres of ctenophores are utilized only for nourishment (as nutritive yolk) and that endoderm is formed later, Chun [72] claimed to demonstrate that the gigantic blastomeres are identical with endoderm, produced by a process resembling regular gastrulation and which is, in fact, gastrulation. In response to Kovalevsky's [73] more recent data, Chun was forced to admit that the "blastopore" corresponds to the aboral pole of the ctenophore's body; without doubt, this is a most extraordinary exception to a widely accepted rule.

Both authors, however, are of the same opinion on the origin of mesoderm. In his *Entwicklungsgeschichte der Rippenquallen* [74], Kovalevsky initially insisted that the star-shaped cells of the gelatinous mass are created from ectoderm by separation both from the exterior integument of the body and from the stomach wall; they later develop pseudopodia which penetrate the mucous mesenchyme.

In his next paper on ctenophore development, Kovalevsky [75] reiterated the same idea in the following words: "Between ecto- and endoderm, a plentiful gelatinous substance is formed; the ectoderm inculcates this substance through cells which acquire the form of star-shaped connective tissue cells." This, however, applies only to *Escholtzia (Callianira) bialata* and *Eucharis*. He admits that "*Beroe* does not seem to form an intermediate gelatinous layer, although this [gelatinous] substance appears to develop directly from the yolk spheres."

According to Chun [76], "a layer is secreted between the ectoderm and endoderm; endoderm cells from the stomach soon start to penetrate it, branching considerably." The invasion of masses of endoderm cells into this layer is not confined to the embryonic stage; in some ctenophores, like *Eucharis* and *Cestus*, it takes place during the entire life-cycle [77].

My own observations were predominantly on *Callianira bialata*, the embryology of which I studied at Messina in 1883 and at Villefranche this spring (1885). In addition, I studied the formation of endoderm in *Beroe* at Naples in 1880, and in *Cydippe* at Odessa in 1882. I shall describe cleavage stages only to the extent necessary to give a very general idea of gastrulation. Since studies on living specimens were not very productive, I worked with embryos treated with a 2% solution of acetic acid or, better, with a mixture of this solution and several drops of 0.5% solution of osmic acid. Let us now turn to observations on *Callianira bialata*.

As is well known, three divisions in a vertical plane lead to the formation of the eight-celled stage, with blastomeres all of the same size (Dr. 1). Equatorial cleavage then occurs (Dr. 2), resulting in the appearance of eight micromeres and eight

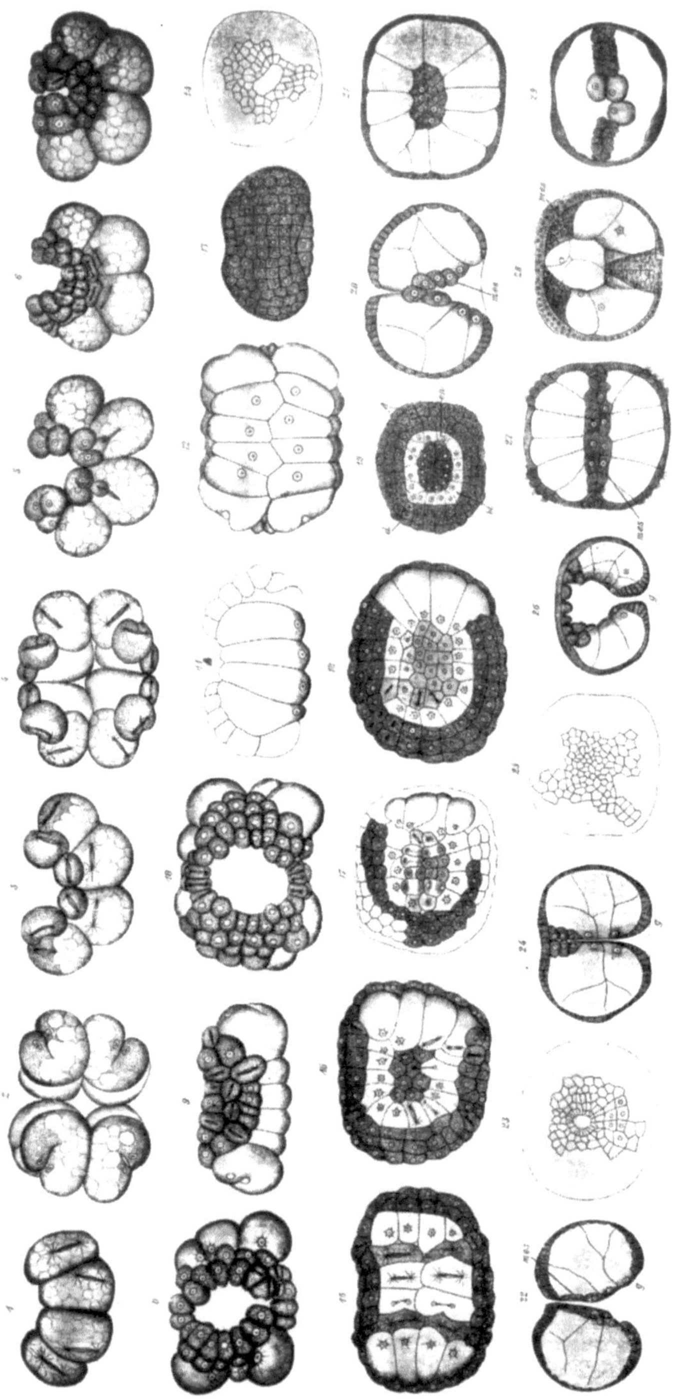

Figure XVIII

FIGURES XVIII–XX

Dr. 1–33 relate to *Callianira bialata.*

Dr. 34–43 relate to *Beroe Forskalii* and *B.ovata*

Dr. 44–56 relate to *Astropecten pentacanthus.*

Dr. 57–76 relate to *Strongylocentrotus lividus* and partially to *Sphaerechimus granularis.*

All contours are made with a prism of Nashe.

Dr. 1. 8-cell stage of division from medial surface. Ocular 2+, Garnak's system 7.

Dr. 2. Transition stage to 16-cell stage. The same magnification.

Dr. 3 and 4. 16-cell embryo with all cells under division. The same magnification.

Dr. 5. The following stage with 20 ectodermic cells. The same magnification.

Dr. 6. The following stage with approximately 48 cells of ectoderm. The same magnification.

Dr. 7 and 8. Similar stage with initiated division of macromers. 2+7.

Dr. 9. The following stage from medial surface. 2+7.

Dr. 10 The same embryo from above. 2+7.

Dr. 11 The same embryo in optic frontal section. 2+7.

Dr. 12. The same embryo from above. 2+7.

Dr. 13. Evolution of embryo with creation of mesoderm. 2+7, height 65 mm.

Dr. 14. The same embryo from the from below. The same magnification.

Dr. 15. Embryo at the beginning of mesoderm formation, from below. 2+7.

Dr. 16. The following stage of mesoderm formation.

Dr. 17. Embryo with mesodermic cells proliferating through division. 2+7.

Dr. 18. The following stage of mesoderm formation.

Dr. 19. Embryo with a 2-cell mesodermic blade. *Bl* – blastopor; *d,d* – dividing cells of ectoderm; *en* – cells of ectoderm. 2+7, height 65 mm.

Dr. 20. Stage of invagination. 2+7.

Dr. 21. The same embryo from below. 2+7.

Dr. 22. The following stage of invagination, *mes.* – mesoderm. 2+7.

Dr. 23. The same embryo from above with pseudoblastopore. 2+7.

Dr. 24. The same embryo in saggital section; *g* – concaving of ectoderm. 2+7.

Dr. 25. The following stage with blastopore on top of it. 2+7.

Dr. 26. The same stage in frontal section; *g* – stomodeum, 2+7, height 65 mm.

Dr. 27. The same stage with cilia in initial formation; *mes.* – stripes of mesoderm. 2+7.

Dr. 28. The following stage in optic frontal section; *mes.* – mesoderm. 2+7.

Dr. 29. The same embryo from above. 2+7.

Dr. 30. The following stage. 2+7.

Dr. 31. The later stage with a cross-shaped mesoderm. 2+7; *n* – wandering cells.

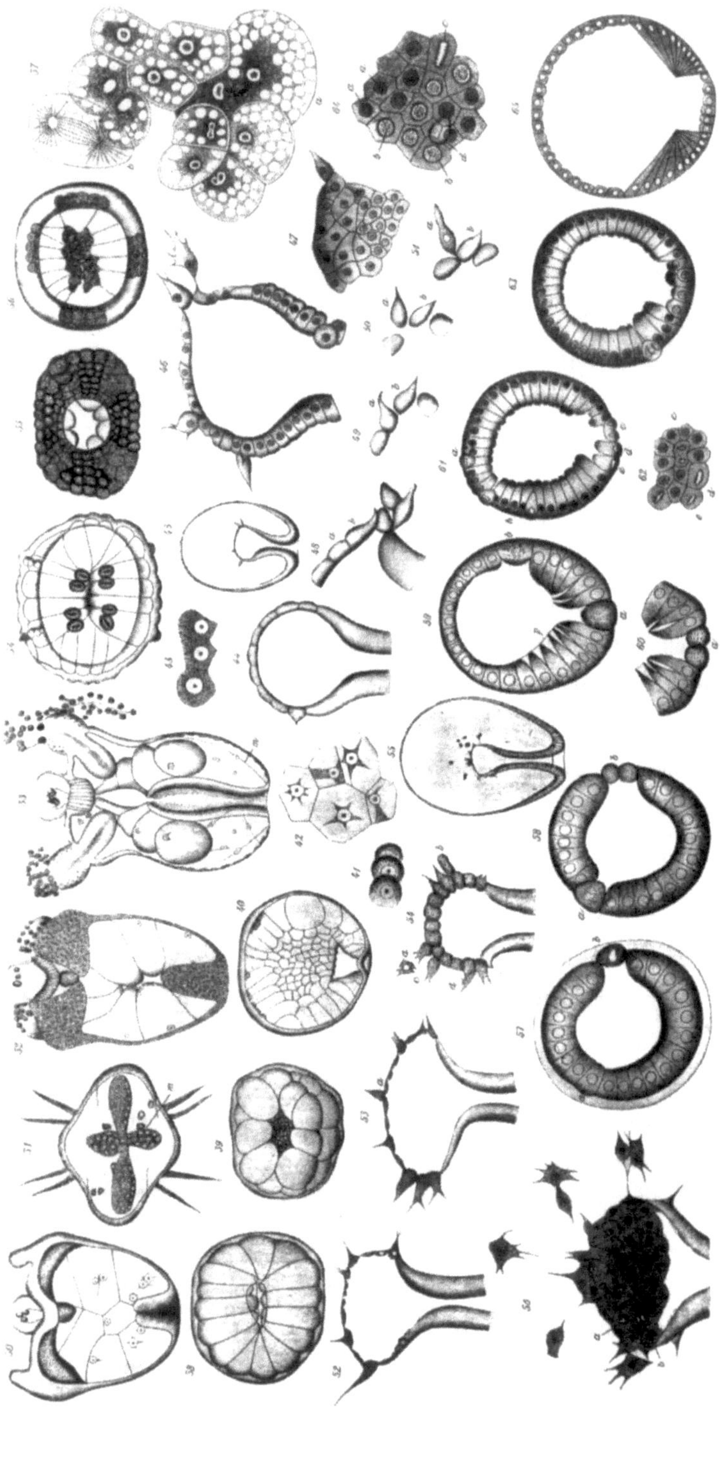

Figure XIX

Dr. 32. The same stage in optic frontal section. 2+7.

Dr. 33. Completed embryonic stage; *m* – significantly contracted muscle fibres. 2+7.

Dr. 34. Embryo *Beroe* with initial formation of mesoderm. A live animal is depicted. 3+2.

Dr. 35. Aboral surface of the following stage. 3+2. Preparation is treated with the acetic acid.

Dr. 36. Similar stage from oral surface. 3+2. A live animal is depicted.

Dr. 37. Part of mesoderm of the same embryo; *a* – budding micromere; *b* – dividing cell of mesoderm. 3+5.

Dr. 38. Stage of concaving of gastral surface. A live animal is depicted. 3+2.

Dr. 39. The following stage from above.

Dr. 40. The following stage in optic frontal section.

Dr. 41. Three cells of ectoderm with cilia. Similar stage. 3+8.

Dr. 42. Cells of ectoderm of the same stage. 3+8.

Dr. 43. Cells of mesoderm of the same stage. 3+8.

Dr. 44 Primary intestine of larva *Astropecten pentacanthus.* 3+7.

Dr. 45. The following formation of larva from plankton. 3+4. Height 65 mm.

Dr. 46. Blind cavity of primary intestine of the same larva after treating by 0.5% osmic acid, dying by a mix of picro- and boraxcarmine, and conservation in glycerin. 3+9, height 55 mm.

Dr. 47. Part of a surface of the same blind cavity (the same magnification).

Dr. 48–51. Four stages of the blind cavity where *a* and *b* represent the same cells. It took half an hour to start (Dr. 48) and to finish (Dr. 51).

Dr. 52. Blind cavity of larva *A. pentacanthus*, grown by artificial fertilization. A live animal is depicted. 3+2

Dr. 53. The same blind cavity two hours later. 2+7.

Dr. 54. The same blind cavity half an hour after treatment by 0.5% osmic acid; *a,b,c,d* – detached wandering cells.

Dr. 55. The following stage of the same series with 15–16 detached wandering cells. 0.5% osmium acid. Picrocarmine+bile glycerin. 3+4.

Dr. 56. Blind cavity with several detached wandering cells of the same larva. 2+9.

Dr. 57. Blastula *Strongycentrotus*. A live animal is depicted. *A,b,* – cells ready for division. 2+9.

Dr. 58. The same blastula 10 minutes later. *A, b* – the same cells as in Dr. 57.

Dr. 59. The following formation of blastula (cilia are not shown). *A, b* – cells with division of nucleus, *p* – protoplasmic membrane.

Dr. 60. Lower part of the same embryo 10 minutes later. Cell *a* is almost completely divided. 2+9.

Dr. 61. Younger blastula with several dividing cells. *A, b* – cells ready for division but not connected to any canal; *c,d,e* – three cells already under division. Osmium fumes. Bilev carmin. Glycerin. 2+9.

Dr. 62. Lower cells of the same embryo from a surface. *C,d,e* – the same cells as depicted in Dr. 61. 3+9.

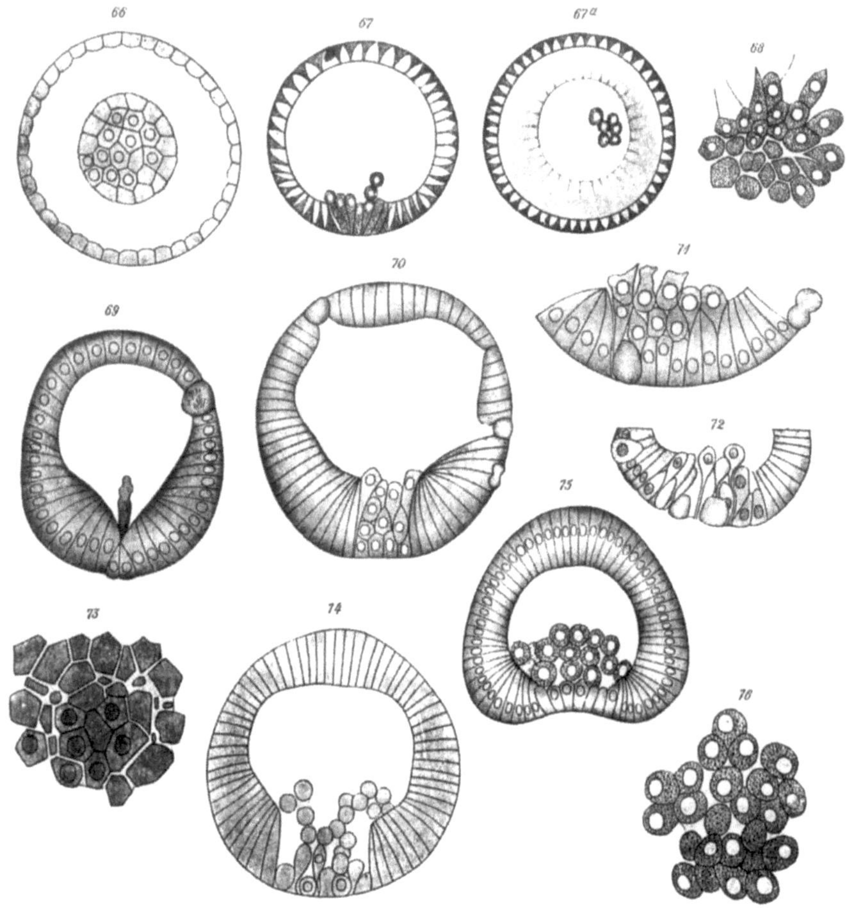

Figure XX

Dr. 63. Similar blastula. Same treatment as in Dr. 61. 2+9.

Dr. 64. Lower cells of the same blastula from the surface. *A* – cells with immobile nucleus; – cells with nucleus ready for division; *c* – cell with elongated nucleus. 4+9.

Dr. 65. Blastula with large deepening on its lower end. A live animal is depicted. 3+7.

Dr. 66. The same blastula from below. 3+7.

Dr. 67. Blastula *Spaerechinus granularis* when first 7 mesodermic cells are coming out. Preparation is treated with vinegar acid. 2+7

Dr. 67A. The same blastula from above. 2+7.

Dr. 68. Lower cells of blastula of the same *Echinus* before their detachment. Preparation is treated with vinegar acid. 3+9.

Dr. 69. Blastula *Strongylocentrotus lividus* with only one wandering cell. A live animal is depicted. 2+9.

Dr. 70. More mature stage of the same species. A live animal is depicted. 2+9.

Dr. 71. Lower part of the same stage. A live animal is depicted. 2+9.

Dr. 72 Lower part of similar stage. Osmium fumes. Bilev carmin, glycerin. 2+9.

Dr. 73. The same from below. 2+9.

Dr. 74. Larva with very intense mesodermic formation. Osmium fumes. 2+9.

Dr. 75. Larva at the beginning of gastrulation. Osmium fumes. 2+9.

Dr. 78. A group of wandering cells of larva. Osmium fumes. Bilev carmin. 3+9.

Dr. 45,55,59–75 do not depict cilia.

macromeres. Strictly speaking, these latter [macromeres] may be regarded as the first ectoderm cells, since they continue to bud off more ectoderm cells (Drs. 4, 5). In addition, ectoderm elements increase in number by the division of micromeres; this process is more regular initially than in the later stages. Firstly, the eight micromeres divide, almost simultaneously, into 16 new ones, and these latter continue to multiply more or less regularly (Drs. 6–8). Their division is multidimensional, which is reflected most clearly by the orientation of the nuclear spindles: for instance, the older ones divide horizontally (Dr. 8), the younger ones meridionally. In spite of these differences, however, their subsequent development is quite uniform, as reflected by the regular, basket-like shape of the ectoderm (Drs. 6–8).

After the total number of micromeres reaches 30–50, budding stops and is eventually replaced by longitudinal division. Of the initial eight, four medium macromeres divide first (Drs. 6, 7). This produces a stage with 12 macromeres of different sizes (see Kovalevsky's text, quoted above, Tabl. VII, Dr. 8), then the lateral macromeres also start to multiply (Drs. 9, 12). Thus a stage with 16 macromeres, interesting in many respects, is created. The ectoderm has a marked tetraradial design, reflected in the cross-like shape of the blastopore aperture and in the development of four ectodermal blades (Drs. 9, 10). These later spread downward and gradually enclose the giant cells. The macromeres, previously also basket-shaped (Drs. 1–8), now flatten considerably (as was previously noted by Kovalevsky). After the completion of the initial longitudinal division, the nuclei migrate from the upper poles of the cells (Dr. 6) downward (Dr. 11); This process can also be seen in the eight medium macromeres and marks the first stage of mesoderm formation.

When the ectoderm that surrounds the embryo approaches the lower edge (Dr. 14), new budding starts on the lower surface; this budding starts in the macromeres and leads to the formation of mesoderm. Eight medium macromeres acquire [nuclear] spindles (Dr. 15), while the nuclei of the eight lateral macromeres remain undivided for quite some time. Budding occurs in an oblique plane, and leads to the formation of eight cells, which occupy the central part of the lower embryo (Dr. 16).

The first eight mesoderm cells are barely formed (one such cell is depicted in Dr. 16, still connected by a ligament to its mother cell) when budding of two of the four lateral macromeres starts again (Dr. 16); this leads to the creation of new mesoderm elements. The other four cells follow their example. Stages with 8, 12 and 16 mesomeres appear, created from the corresponding numbers of macromeres. The regularity of this process, however, is often broken by the first mesoderm cells starting to divide before the rest of the mesomeres are formed (Dr. 17); this leads to a premature increase in number. In other cases, by contrast, the formation of mesoderm is interrupted at an earlier stage, so that we can observe a plate with only 12 cells (Drs. 19, 21). With these exceptions related to the number of cells and the moment of their initial differentiation, we can note the following rule: in *Callianira*, mesoderm appears as a result of budding on the lower poles of macromeres; this tissue forms a plate, consisting of several cells, and placed horizontally, *i.e.* directed to the frontal region.

Soon after embryonic mesoderm is created, invagination of the macromeres starts; these macromeres also form mesoderm, which is pulled inside the embryo.

The ectoderm now consists of a single layer of comparatively small cells, which still retains a small opening on the upper surface of the embryo (Drs. 20, 22, 23), while at the lower pole, by contrast, they surround a gigantic blastopore (Drs. 19, bl, 20). Along the meridians, which are later transformed into ctene, rapid cell division may be seen (Dr. 19, d). Within the invagination, mesoderm cells penetrate into the cavity of the gastrula (Dr. 20, mes), where they remain unchanged for a while, retaining a sac-like shape. Gradually, however, they move to the upper surface of the embryo and develop a plate-like form (Dr. 24).

The endoderm was created, as shown earlier, from 16 micromeres. The same number of endoderm cells is retained for a long time; their number increases only after subsequent invagination, but still not very substantially.

When invagination is complete, the upper opening completely disappears, filled by smaller ectoderm cells (Drs. 23, 25). At the same time, secondary invagination of the ectoderm occurs on the lower surface of the embryo (Drs. 22, 24, 26). This invagination, as is well known, creates the so-called stomach or stomodeum.

If gastrulation is observed in the described sequence, there is no doubt that the blastopore is formed on the oral surface of the embryo, *i.e.* that the long-lasting aperture at the upper pole is, in fact, the pseudoblastopore.[31] This orientation was also confirmed by successive drawings of the embryo with a *camera lucida*;[32] this was particularly successful in the case of the large eggs of *Beroe*.

After reaching the upper surface of the embryo, the mesodermal plate, although still directed to the frontal region (Dr. 26), gradually separates from the embryonic endoderm. At the start, the cells are comparatively large, and, therefore, not numerous (Dr. 27), but they soon start rapid division, particularly on the bilateral surfaces, where two bumps form (Dr. 28 mes). This is the period when the mesodermal plate is divided into right and left halves (Dr. 29), between which a significant aperture is gradually revealed; this aperture, however, is soon filled by two growths of mesoderm (Drs. 30, 31). Thus a cross-like mesoderm appears; its lateral parts produce tentacular mesoderm, while, in my opinion, both medial growths become the sources of wandering cells.

This conclusion follows from the fact that these cells are found next to the embryonic mesoderm (Dr. 31, m), and that newly-formed wandering cells are identical to elements of these embryonic cells in size and other parameters. I could not trace the formation of wandering cells in living specimens because the embryo is very active at this stage; it therefore should be immobilized by a decisive lethal action. Stopping movement in a live embryo is possible merely by rupturing the egg membrane; without such a membrane, the embryo cannot stay alive in seawater.

In proposing, on the one hand, that wandering cells are derived from medial embryonic mesoderm, I cannot, on the other hand, categorically deny that lateral stripes of mesoderm may play a part in the formation of these cells. Nevertheless, I have to emphasize that ectoderm, while still in its embryonic form, does not create wandering cells. Each time I tried to verify this suggestion, I got a negative result. Judging from their drawings, Kovalevsky and Chun also did not observe transition of ectoderm cells to wandering cells. Their first depiction of wandering cells was always in the middle or lower part of the embryo, whereas these cells initially appear somewhere close to the upper pole and go down only later. When Chun claimed that some ectoderm cells, in their capacity as muscle cells, change in adult ctenophores into a gelatinous mass, this is not necessarily evidence in favor of a second way of mesoderm formation, since, according to R. Hertwig [78], the above mentioned muscles are endodermal and penetrate the mesoderm later, as is generally observed in many organs.

In the later stages of the embryonic development of *Callianira*, lateral parts of the endoderm increase considerably in size, while their elements [cells] become much smaller and less distinct (Dr. 32). Their median parts gradually disappear, perhaps due to removal by wandering cells. Laterally, connective mesoderm cells link the embryonic ectoderm of the tentacles and then form muscle fibers (Dr. 33). Further histological processes were, unfortunately, too difficult to observe. At the same time, the number of wandering cells increases; a significant number of them occupy an empty space in the lower part of the embryo and some of them change into muscle cells.

What I observed in the above-mentioned ctenophores, seems sufficient to postulate that the facts of mesoderm formation in *Callianira* represent features common to all ctenophores. Thus, for example, I saw exactly the same mesodermal budding of macromeres in *Cydippe* from the Black Sea, with a completely analogous process of gastrulation.

Beroe turns out to be much more interesting because it has no tentacles; thus mesoderm formation cannot be linked to the formation of tentacles. Yet we see much the same picture, namely that small cells are formed by budding from the lower poles of macromeres to create a lamellar mesoderm. In the earliest stage observed in live specimens, I saw eight symmetrically positioned mesoderm cells (Dr. 34) which then immediately divided into 15 cells. There is no doubt that these eight cells were the offspring of neighboring macromeres, since budding of macromeres can still be observed in later stages with a significant number of mesoderm cells (Dr. 37, a). Mesoderm cells themselves continue to divide, so the mesodermal plate grows very fast. Further, macromeres are being surrounded by more and more ectoderm cells, while the blastopore, which formerly occupied almost the entire lower surface of the embryo (Dr. 34), becomes much smaller, nevertheless still retaining its oval shape (Dr. 36).

Invagination itself starts after the last of the above-mentioned stages is completed; it corresponds to the same process in *Callianira*. In contrast to the above data, the surface of the gastrula is here sparsely surrounded by ectoderm cells, not to the same significant degree as in other ctenophores. The edges of the blastopore continue to frame an aperture in the shape of a more or less regular oval (Dr. 38). The in-growing mesoderm becomes more and more invaginated until it becomes invisible from the surface of the gastrula (Dr. 38); to see it now, the embryo should be viewed from the side [in optical section]. Such an observation reveals a plate of mesoderm covered by ectoderm and consisting of even more numerous cells (Dr. 39). The pseudoblastopore on the aboral surface, quite large before invagination (Dr. 35), becomes overgrown in this species [*Beroe*] even earlier than in *Callianira*; as in the latter it has nothing to do with gastrulation. In *Beroe*, the entire process of invagination can conveniently be observed in the same embryo; so this description of the structure of the embryo seems incontrovertible, especially as this description agrees with that given in the second article by Kovalevsky, and does not correspond to Chun's description.

In the later stages, when ectodermal thickenings arise on both sides of the embryo (these thickenings can be considered as rudimentary embryonic tentacles [Dr. 40]), the mesoderm is divided into two parts (in the same manner as in *Callianira*). A secondary invagination, that of the stomodeum at the lower pole of the embryo, takes place at the same time, and the ctenophore's ectoderm produces several short [rows of] cilia (Dr. 42], the first embryonic vibrating plates. The endoderm, at this stage, is represented by parenchymatous tissue consisting of expanded cells (Dr. 42); the mesoderm cells, by contrast, are more dense, granular, and contain far less fluid. When the embryo becomes motile, the mesoderm cells, so small with respect to the whole body, become increasingly difficult to observe; that is why, regretfully, I can give no further information on the development of this embryonic layer. In addition, I did not have a recent opportunity to re-examine it.

Previously, only Al[exander] Agassiz observed and illustrated mesoderm cells of ctenophores, but he did not ascribe any significance to them. Some of his drawings of the embryo of *Idya rosea* depict a number of remarkable cells on the aboral side of the body; I consider these cells to be nothing other than embryonic mesoderm.

We can conclude from the foregoing facts that a ctenophore gastrula is the result of invagination and is formed after the completion of an earlier overgrowth of endoderm cells; it is, coincidentally, an embolic and an epibolic gastrula. The design of the latter is regular, while the blastopore, which then transforms into the inner aperture of the gastrula, is an oral blastopore. Since the production of endoderm does not originate in the animal pole itself but in a ring-like embryo, what is created here should not be regarded as a genuine blastopore but as an inner pseudoblastopore; this latter is eventually overgrown and becomes the rudiment of a sense organ.

Ctenophores are the only Metazoa in which mesoderm is formed during embryonic development as a separate embryonic layer. While mesogloea in medusae and polyps is formed in the post-embryonic stages of development and even then does not acquire a separate entity, mesoderm in ctenophores is formed very early and represents an integral embryonic formation. It must be regarded as a genuine embryonic layer, *i.e.* mesoderm, although, according to the Hertwig brothers, this is more than doubtful. Accepting their criteria, we should consider the mesoderm of ctenophores as an embryonic layer because it consists of "embryonic cells combined into an epithelial plate" [79], but never as mesoblast, since, firstly, it is of earlier origin, and, secondly, it is formed differently from "a convexity or fold in the endoblast" [80]. The facts of mesoderm [formation] in ctenophores cannot be reconciled with the notion of embryonic mesenchyme (as also proposed by the Hertwig brothers), both because it is not created from "embryonic cells, separated one by one from an epithelium," and because embryonic mesenchyme does not have the capability to form an embryonic layer.

If we thus ascertain that the particular mesoderm of ctenophores is not explained by the Hertwigs' theory of a middle embryonic layer, we obtain another proof that this theory cannot sustain empirical verification. The middle embryonic layer is too variable to be presented as just two types; furthermore, everything in its development is so intertwined that these two types cannot even be clearly distinguished from each other.

I should like to add one more consideration to those presented in the third part of these studies, a consideration that shows the close connection between mesenchyme and the Hertwigs' mesoblast. As I mentioned elsewhere [81], some peritoneal cells of *Nais proboscidea* become disconnected and, while developing amoeboid movements, start to overgrow different foreign bodies which have penetrated the body cavity, like, for example, a young larva of *Gordius*. Thus we see that cells of peritoneal epithelium derived from mesoblast, even in adult Annelida, can transform, under the influence of strong irritation, into mesenchyme cells. An analogous fact was observed long ago in mammals by Ranvie [82]: some peritoneal cells of an inflamed peritoneum acquire the ability to capture foreign bodies and devour them by producing amoeboid outgrowths. After that, is it still possible to talk about the "total difference" between mesenchyme and mesoderm?

COMPARATIVE EMBRYOLOGICAL STUDIES V.
V. ON THE CREATION OF WANDERING CELLS IN ASTERIDA AND ECHINIDA.[33]

After Genzen [83] discovered in 1863 what he called a "secretory" tissue in a young brachiolaria[34] from Keil Bay, and established that this tissue creates several pine-shaped cells that eventually separate from the tissue and move into an intermediate galatinous substance, this concept became quite popular and was confirmed by different investigations. Gotthe [84] and I [85] pointed to the similarity of the occurrence of wandering cells in bipinnaria[15] larvae. The same occurrence was discovered by Zelenka [86] in *Holothuria*. He reported the following with regard to *Holothuria tubulosa*: "Twenty two hours after fertilization is completed, several cells (their number varies from four to ten) leave a thickened area and create a cellular complex which, from now on, is the only source of formation of mesoderm cells ...." Many scholars shared the opinion that initially wandering cells either disperse in a random manner or form an unpaired aggregation, and, in either case, do not show a symmetrical arrangement.

This part of the problem, however, was interpreted differently after Hatschek [87], in 1878, who emphasized a close link between the Echinodermata and the Bilateria and attempted to re-examine the development of mesoderm. In his theoretical review, he said, "The formation of mesoderm in the Echinodermata starts in the area of the embryonic bubble [blastula] where detachment of mesodermal cells in other Bilateria takes place, but no bilateral location of embryonic mesoderm has ever been observed in the Echinodermata."

This problem still awaits further clarification. Zelenka's text on embryonic layers in *Echinus* appeared a year later [88], and described embryonic mesoderm as arranged symmetrically from the very beginning. "The extreme transparency of the larva of *Echinus miliaris* and *Toxopneustes brevispinosus* allows us to trace the formation of mesoderm in every detail. Within the central area of thickening of the blastoderm, which coincides with the longitudinal axis of the later gastrula, a funnel-shaped depression is created; in approximately an hour, it extends into a fissure-like groove, which disappears only when the embryonic mesoderm is formed. As a result of this groove, bilateral symmetry of the larval body is established; it develops by the formation of embryonic mesoderm on both sides of the groove by the division of a small number of endoderm cells grouped into two piles of 4–6 cells, but sometimes just two [cells]. These cells soon migrate from the original site and penetrate the gelatinous substance; they move like amoeboid cells and undergo multiple division."

Hatschek [89] later undertook his own research, and, although he too observed symmetrical embryonic mesoderm in *Toxopneustes lividus*, he described it in a completely different way from Zelenka. According to Hatschek, mesoderm in this echinoderm "arises from just two cells which touch each other in the median line on the endodermal axis of the blastula; while dividing, these cells move into the cleavage cavity."

Having studied gastrulation in the Echinida, I had to consider, in passing, endoderm formation in these animals [90]. Although I studied several embryonic series in *Echinus microtuberculatus*, I could not confirm Zelenka's data. In both the time of origin of endoderm cells within the blastoderm and later, after their penetration of the cleavage cavity, I saw no notable symmetry. Similarly, my studies did not confirm a link between the depression described by Zelenka and a symmetrical arrangement of the initial elements of mesoderm.

In his next work on embryonic layers in the Echinodermata, where prominence is given to the history of development of the wandering cells, Zelenka [91] once more changes his opinion. He thus completely abandoned his earlier ideas on asymmetrical embryonic mesoderm in *Holothuria*, substituting a paired arrangement of embryonic mesoderm from two initial cells, completely adopting the views of Hatschek, whose original sketches he placed next to his own. According to Zelenka's revised opinion, mesenchyme in the Echinida is created from two initial cells, the division of which forms two symmetrically arranged stripes of mesenchyme. Mesenchyme cells divide and move randomly into the cavity of the blastula.

Zelenka [92] extended his observations to other Echinodermata (*Synapta*[35] and *Ophioglypta*[36]) and finally concluded: "In the Echinodermata, mesoblast is formed in two ways: a) from two initial 'mesenchyme' cells, and b) from folds in the archenteron. The two initial mesenchyme cells give rise to two stripes of mesenchyme, which, together with their branching cells (the wandering cells), penetrate the blastocoel (the cavity of the blastula) and finally create two sorts of tissue: circular muscle tissue and cutis. *Synapta* does not form two types of mesoderm because, unlike all other Echinodermata, gastrulation here precedes mesoderm formation."

Since I carried out my first observations on *Echinus microtuberculatus* only, I considered it highly desirable to revise this issue with respect to other Echinodermata; also, the question of the origin of the wandering cells became especially interesting after I found that these elements play a significant role as phagocytes. That is why I undertook a detailed analysis of embryos of *Sphaerechinus granularis* in Messina, in 1883. I was not only able to confirm my previous data but to move further, observing the creation of wandering cells from migrating pear-shaped cells from embryonic ectomesoderm.

When Zelenka's paper, mentioned above, was published several months later, I noticed regretfully that our independent studies were in total disagreement. I found many of Zelenka's data not to be particularly convincing, especially as he never drew the full length of a mesenchyme strip, which is crucial. The reason for the discrepancies, however, remained unclear to me. In such a situation, it seemed vital to renew my studies. In order to stimulate the situation (especially as I was not going to carry out marine work in the near future), I published a short note on my research [93]. A categorical rejection from Zelenka [94] followed soon after; here he reiterated his observations and insisted on their complete accuracy, without, however, presenting any new argument. To solve this problem, I had to return to my research. At the end of last year, I studied relevant stages of *Strongylocentrotus lividus* at Tangier [95], and this spring, the embryology of Echinodermata at Villefranche. In addition, I analyzed the development of *Astropecten auranciacus* and *A. pentacanthus* this May, at the Zoological Station in Trieste. I am very grateful to Prof. Klaus and Inspector Greffe for their kind consent to use the facilities of this station.

I shall start the presentation of my results with *Astropecten*, which is much more suitable for discussion of the problem than the Echinida. Young bipinnarias of *Astropecten* can easily be obtained from pelagic plankton or by artificial fertilization. *Astropecten auranciacus*, however, was less suitable for research than the smaller *A. pentacanthus,* which becomes adult in May. Since larvae of both animals are colorless and transparent, a lot can be seen in live specimens; this is a huge advantage, especially where moving cells are concerned. For a more detailed study, however, analysis of preserved larvae is indispensable. To attain this goal I prepared larvae [by adding] a drop of 0.5% osmic acid (fumes were not very effective in this case), then I stained them with picrocarmine, and finally placed them in glycerin. Preparations in balsam are not as good; sections of larvae of *A. auranciacus* were completely useless because cells in such a thin lamina can easily be displaced and thus lost.

Artificial fertilization of *A. pentacanthus* should preferably be done before midday because wandering cells start to form after 27–30 hours; this stage in *A. auranciacus* lasts much longer, so the same stage is to be expected some 48 hours after the eggs are removed [from the female]. Sperm may be dropped into the vessel containing the eggs; there is no need to wait for a polar body to appear.

When division of the yolk is completed, a characteristic folded blastula forms; it gradually removes the egg membrane and acquires the usual bubble shape. Such folded blastulas, characteristic of both species of *Astropecten*. look very similar to abnormalities found by Folle [96] in *Asterias glacialis*[37] and called "polylastic". The same result can be observed in *Astropecten*: here not merely regular folds but sac-like invaginations result from a process quite similar to gastrulation, although no similarity between these two formations can be noted in either timing or basic structure. Like the folds, these sac-like invaginations gradually flatten out to give the regular bubble-like blastula mentioned above, which grows considerably as this change occurs. Only in exceptional circumstances do abnormal larvae retain some of these sacs, which then elongate and develop interconnections, like canals; thus peculiar bipinnarias are created, with a system of canals opening to the outside through several pores (as in some sponges).

Gastrulation occurs exactly as described more than twenty years ago by Al. Agassiz [97] in *Asterocanthion*. Initially cylindrical cells of the invaginated archenteron flatten on the inner blind end, creating a retort-shaped endoderm, with the bowl consisting of flat epithelial cells and the neck of cylindrical cells. The epithelial cells of the bowl gradually flatten even more; then some of them begin putting out short pseudopodial projections (Dr. 44 [Figure XIX]). At this stage, this phenomenon can be seen in some "bowl" cells which, however, still retain their initial relationship with neighboring cells. If a drop of osmic acid is quickly added to the preparation, the pseudopodia are then withdrawn, and the picture presented is identical with the previous stage. Only a little later do the pseudopodial processes become longer; now, however, the epithelial cells no longer retain their original links to neighboring cells and start to produce the first wandering cells. Usually there are several such cells: four or five or sometimes more simultaneously break free of the bowl, so, in spite of numerous attempts, I was not able to find a stage with two wandering cells.

Once mesoderm begins to form, amoeboid cells can be seen which separate more or less completely from the epithelium; at the same time, other wandering cells, although completely free, attach to the surface of the bowl for quite a while. The nuclei of such cells are, as a rule, inactive; I never witnessed direct formation of [other] wandering cells through division of these inactive cells. Although mesodermal cells may sometimes be connected, through pseudopodia, to epithelial cells on the bowl, the form of the nucleus does not lead to any conclusion on active division.

To obtain reliable results in studying amoeboid elements, it is crucial to trace the entire process in one and the same cell. I succeeded in observing the origin of a wandering cell from such a single cell on several occasions. Drs. 48–51 show four stages in the development of a cell from the bowl; cell "a" separated from its neighbors and transformed its status from an epithelial cell to a wandering cell in just half an hour. This process often lasts longer; I observed several instances of cells with pseudopodia already emerging, but, instead of emerging further, the pseudopodia were retracted, and the cells returned to their previous state.

Drs. 52–54 may be used to understand the progressive detachment of a cell (52 and 53 were sketched from live specimens, 54 from a preserved one which had been treated with 0.5% solution of osmic acid). Dr. 52 shows only those cells with pseudopodia, while Dr. 53, drawn two hours later, shows an optical section of several cells that are almost detached and one completely separated (a). Three more cells soon joined them. While these, the first of the detached cells, progressively distanced themselves from the epithelium of the bowl, the creation of new mesoderm continued (Drs. 55, 56). It can be established, both from live specimens and (better still) from preparations, that the production of new wandering cells occurs mostly as the result of continuous migration, and not from division of those mesoderm cells that are already free. Active division of wandering cells is a rather exceptional method of creating mesoderm, and, as a rule, the nuclei are inactive both in cells that became free earlier and those just emerged.

I observed cell division predominantly in epithelial cells that retained their flat shape (Dr, 56, a, b), and, by multiplying, filled vacant spaces left after the departure of migrated cells. Usually these two processes occur together, so that the epithelial bowl keeps its shape (although significantly flattened). In some cases, however, migration is so rife that a quite obvious gap appears on the upper surface of the bowl. This gap is filled only in the later stages of development. Such abnormal larvae, it seems, finally change into completely normal bipinnarias.

Creation of endoderm continues in practically the same manner only a little slower during the subsequent development of larvae. Migration of new cells stops when the coelomic pouches start to form [98], while epithelial cells continue to put out amoeboid projections long after that.

I observed these processes in larvae of both *A. pentacanthus* and *A. auranciacus*. Results recorded in artificially reared larvae appear to be identical to those in freshly caught bipinnarias. I was also fortunate to raise several specimens, used for my studies on mesoderm, to mature bipinnarias, up to the formation of the starfish rudiment. I mention all of this to eliminate any doubt regarding the normal course of events described here. Observations by me and other researchers on brachiolarias[38] agree, in general, with data from studies on bipinnarias.

As I mentioned elsewhere [99], the larva of *Astropecten* is identical with that described by J. Müller as *Bipinnaria* from Trieste [100]. It is still difficult, however, to determine the species of Müller's *Bipinnaria*, as the larvae of *A. pentacanthus* and *A. auranciacus* strongly resemble each other in size and transparency.

From the Ophiurida, I observed only *Ophiothrix fragilis*, larvae of which I bred after artificial fertilization. Unfortunately they were not very suitable for studies on the formation of mesoderm, as I could not see much. The formation of wandering cells in *Ophiothrix fragilis* appears to be similar to the same process in the Echinida, to the description of which I turn now.

I studied [larvae from] artificially fertilized eggs of *Strongylocentrotus (Toxopneustes) lividus* and *Sphaerechinus granularis*. As for *Echinus microtuberculatus*, I did not renew my studies on it after 1880. However, the formation of wandering cells in all three species is quite similar. In addition to live embryos, I studied those processed by osmic acid fumes and stained with carmine.

In order to explain the formation of wandering cells in the Echinodermata, young blastula stages have to be studied first; particular attention should be paid to cell division. I cannot confirm Zelenka's statement, namely "after segmentation of the egg is completed, any further cell division ceases for quite a while" [101], because I found some cells dividing at all stages (although this number varied in different stages). A cell preparing to divide, in addition to internal changes, transforms its entire shape; its orientation apparently turns (Dr. 57, a), then the entire cell rounds off and significantly shortens (Drs. 5, a; 59, b).

The result of all these changes is a striking difference between our dividing cell and the neighboring cylindrical elements, as well as its displacement to the bottom of the canal-shaped space [between cells]; the nucleus becomes spindle-shaped, and the entire cell then divides along a radius of the blastula (Dr. 5, b). The longer the neighboring cells, the deeper and more obvious the canal containing the cell about to divide. Thus cell "a" in Dr. 59 is more obvious than cell "b", which is at the same stage of development. When a dividing cell is located among cells that are considerably flattened, it may be missed in cursory examination. On the other hand, the eye of the observer is always drawn to a place where several cells are about to start simultaneous division and, consequently, are surrounded by noticeably elongated elements. If a cell which is about to divide is located too close to other cells, we see a canal as in cells "a" and "b" in Dr. 61.

Inert cells in preparations treated as above are uniformly colored (Dr. 64, a), while those ready to divide are amazingly transparent: they look like bubbles filled with fluid and packed with irregularly shaped flakes (Dr. 64, b). They later become even more transparent, while the flakes tend to escape the observer's eye and become even more irregular and longer (Drs. 61, a; 64, c). Nuclei preparing for division are visible even at low magnification, and can easily be observed in this state. The changes in cell shape described here are not specific to Echinidae only; I saw them in the embryo of *Astropecten*, and they are also widespread in medusae. It would be a mistake, however, to regard them as a regular feature of cylindrical blastoderm cells; for example, I did not see either compression or rounding of protoplasm in the division of very extended ectoderm cells in the embryo of *Sagitta*.

After this consideration of some facts of cell division, allow me to discuss stages preceding the formation of wandering cells [102]. Cylindrical blastoderm cells, completely identical to each other at first, change, so that cells that form the upper covering of the larva flatten considerably and acquire a cubical or even more flattened shape, while cells of the lower part of the body become progressively longer. The latter then connect to each other by median projections, so that a very thin protoplasmic covering is formed (Dr. 59, p). This gradually spreads to cells of neighboring parts of the body, sometimes to the upper cells of the ceiling (Dr. 67, a). Thus a very peculiar linkage of cells is formed, observed so often that it cannot be regarded as something accidental or abnormal. Cells about to divide do not participate in this process, nor does the embryonic ectomesoderm; so a thin corolla is created by confluence of the connecting threads on the lower part of the larva (Drs. 59, 65, 67,a). No matter how often this phenomenon is observed, there are still other, completely normal larvae that do not have such a developed linkage of cells.

There are many individual variations in stages of cell division before the creation of wandering cells. Sometimes cell division is concentrated mostly in the lower part of the larva; in many cases, by contrast, it spreads in a regular manner to the entire blastoderm; in yet other cases, although quite rare, it hardly ever occurs in the lower part. When several cells in the lower part of the larva are preparing to divide simultaneously, a significant invagination is formed; the bottom of this invagination is paved with short cells, while the cells making up the sides are often tall and are inactive and linked to each other (Dr. 65). For a more detailed examination of these stages, appropriate larvae should be screened not only in longitudinal optical section but also from above and especially from below. In this latter case (Dr. 66), it is evident that the shortened cells occur, in irregular fashion, next to each other, and that they retain, more or less, their initial position. Connecting links are often destroyed by treatment with osmic acid (Dr. 63), but the structure of the nucleus and the arrangement of cells can be discerned even better (Drs. 61–64). It can be seen that, during these stages, several cells move further into the cleavage cavity (Drs. 63, 64), but still continue to participate in the formation of the wall of the blastoderm.

This phenomenon, as shown by observation of live larvae, depends on the amoeboid movement that takes some cells into the cleavage cavity. While the major part of the contents of the cell, including the nucleus, leaves its previous location, the cell itself is still connected to it by a thin stem (Drs. 67, 70–74); this clearly shows its origin. For a clearer picture, a larva treated with osmic acid should be viewed from below; normal polygonal blastoderm cells are connected by small protoplasmic projections, sometimes polygonal, sometimes rounded, which, under close observation, reveal the fixed end of each stem (Dr. 73). Comparison of several such stages shows that there is no uniformity of structure in cells attached to stems.

For quite a while, I assumed that stimulation to activity in wandering cells required the prior formation of a marked invagination by cell division at the lower end of the body. Later, however, I observed several instances of normal stages where cell division occurred predominantly on the sides of the body (Dr. 70), and where precisely the lower pole was marked by a dense accumulation of wandering cells, in spite of lack of nuclear activity. In some cases, the formation of endoderm is triggered by the migration of just one cell into the cleavage cavity (Dr. 69); usually, however, it occurs when several cells leave the blastoderm as a group (Drs. 70, 71). Such cells, still in groups, will move into the cleavage cavity later. In cases in which wandering cells are produced rapidly, a mass of such cells enters the cleavage cavity almost simultaneously; this creates holes at the bottom of the blastoderm, and gives the appearance of devoting all the cells of the lower part to the formation of mesoderm. Further observation, however, reveals that several cells at the very bottom remain there and change their shape to create invaginated elements of embryonic endoderm (Drs. 74, 75).

Both at times of migration and after, wandering cells of embryonic mesoderm form an irregular group or big mass which is not, as Zelenka claimed,arranged symmetrically. My Dr. 76 depicts wandering cells of the larva of *Sphaerechinus* viewed from above; the arrangement of cells together with the inactivity of the majority of the nuclei is quite apparent. My observations indicate that increase in the number of wandering cells is mostly the result of immigration of new elements.

Summarizing my data on the formation of wandering cells in the Echinida, this process may be regarded as an invasion of cells with their entire content from the lower part of the blastoderm into the cleavage cavity. There is, apparently, no symmetry or particular regularity of arrangement in the elements of blastoderm. Before cells leave the mesoderm, or during this process, cell division occurs to compensate for the loss of blastoderm cells (only incidentally linked to the formation of mesoderm). This process differs from that seen in the Asterida only in some unimportant details. Thus, for example, detachment of cells from the blastoderm in the Echinida occurs earlier than in the Asterida; in the latter they derive from flattened cells at the blind end of the invaginal sac, while the cylindrical or cubical cells in the Echinida detach from adjoined embryonic endomesoderm. This early differentiation is perhaps linked to the skeleton-forming role of the wandering cells, since it is also reported in larvae of Ophiurida.

The described development of wandering cells in the Echinida agrees with my previous data, including that obtained from studies on the Asterida, but does not confirm Zelenka's views. Zelenka'a report and drawings group together all the observations on stages in cell division and formation of mesoderm. Shortened cells with big grainy nuclei (Drs. 25, 46–48, m) should be regarded, according to the foregoing, as cells ready to divide, but these divisions, as I have just shown, vary considerably. It is quite possible to encounter instances in which a cell has just divided into two new ones (Dr. 60, a), as well as others where two cells are ready to divide simultaneously (as, for example, in Zelenka's Drs. 22, 23, 46). In both cases, division has resulted in two obvious cells located at the lower pole; further consideration teaches us not to rush to any conclusion on the creation of two-celled embryonic mesoderm. Such a larva with two cells nevertheless exists, but the same cannot be claimed about stages with two stripes of mesenchyme.

No matter how often I attempted to analyze this problem and how diverse the particular grouping of cells preparing for division, I never found (viewing the larvae from below) any grouping of these cells in a stripe. In addition to specimens depicted in Drs. 62, 64, 66, 68 and 73, I drew several others from life using a prism, and I never encountered any regular arrangement, in stripes or anything else. Zelenka, also, did not provide a drawing in which mesenchyme stripes were visible; his Dr. 47 and 48 merely show two longitudinal sections of the supposed stripes, while Dr. 26, m depicts "all the cells which create both stripes of mesoderm" and thus schematizes the entire picture. The only convincing proof of possible mesenchyme stripes would have been in observation from below; such a view is, however, completely missing from Zelenka's drawings.

Zelenka does not consider the role of the shortened cells in cell division. Among his hundreds of drawings of cells with nuclei, there is not even one that is actually dividing. When he talked of the division of primary mesenchyme cells (as, for instance, on pp. 44–45), it is deductive reasoning, not based on any firsthand data. (It has recently become much easier to obtain such data, thanks to detailed studies on changes in the nucleus.) In general, Zelenka seems not to have observed the formation of wandering cells from the pear-shaped, and later stem-like, blastoderm cells; his drawings verify that he missed this stage. And finally, the gap is too big between his Dr. 48, depicting the so-called mesenchyme stripe consisting of ten big cells, and Dr. 49, where we see about 30 small pill-shaped cells clearly separated from each other and already in the cleavage cavity of the blastoderm. These stages cannot be connected by simple deduction, as he tried to do.

If Zelenka's data is unreliable on the formation of mesenchyme in the Echinida, species of which he studied at considerable length, this applies even more to his data on other Echinodermata. Thus, for example, he claimed that "the creation of both primary mesenchyme cells in the Ophiurida occurs in the same manner as in the Echinida (Dr. 45)", and he refers to his Dr. 63 where, however, only one rounded cell "m" is free in the cleavage cavity with two cylindrical protruding cells, *i.e.* a total of three wandering cells are visible. The inert nuclei depicted in these two cells argues against the suggestion that this other cell was formed as a result of division of one of them, but [its formation was] independent of neighboring cells.

Zelenka interpreted his findings on the larvae of *Synapta* in the context of his two-cell theory. Thus, for example, he considered two cells at the upper end of the archenteron as nothing less than "two cells of primary mesenchyme". Zelenka claimed to have encountered such larvae, with two isolated mesenchyme cells, quite often; he assumed that further division of the two primary cells takes place in only a few hours after their complete separation from the archenteron [103]. His claim that the first two wandering cells are primary mesenchyme cells that subsequently divide, he considered as self-explanatory, with no need of experimental verification. Paradoxically, even his drawings testify against this (for example, his illustration of a larva in Dr. 83, with five mesenchyme cells sketched as indistinct spots, suggests conclusions other than those that he actually drew).

The same can be said about his data, especially his observations on *Synapta*, which do not correspond to his theoretical views. He said, for example: "Larvae in which the free end of the archenteron has an irregular outline and also others with star-shaped cells can be encountered quite often; I consider such a formation to be pathological, as is also the rapid division of mesenchyme cells." [104]. Such larvae with numerous wandering cells are very similar to bipinnarias described above; the production of such cells in auricularias[39] strongly resembles the corresponding process in the Asterida. If Zelenka regards these stages as pathological, he should prove it, and this does not seem to be particularly difficult. But the very fact that there were several such stages among young auricularias caught in tow-nets contradicts Zelenka's claim.

Wandering cells in the Echinodermata, as far as can be observed in studies on Asterida and Echinida, detach either from endoderm or from fragments of blastoderm and then move into the cleavage cavity; their functions in the cavity are quite varied. This developmental process is quite similar to that which can be seen in adult sponges, where mesoderm cells detach from endoderm. It resembles also, as far as I know, the process of mesoderm formation in *Rhopalonema*[40] and other Acalepha. The formation of wandering cells in the Echinodermata is thus of considerably lower status on the evolutionary scale than the formation of mesoderm in the Ctenophora, described above.

Reviewing the data on the formation of mesoderm in lower multicellulars (Metazoa), such as sponges and Echinodermata, convinces me that the two-cell theory is not applicable here. Since this theory is more appropriate to some higher multicellulars, it may be assumed that a two-celled embryonic mesoderm portrays an early stage of differentiation; yet there are cases where mesoderm is represented by a single cell. So, while the same concentration, together with precocious differentiation, should be considered an attribute of embryonic development acquired later, creation by detachment of wandering amoeboid cells must be viewed as a rather primitive process. This relates to the role of the wandering cells which regularly perform as phagocytes, *i.e.* function in a manner so apparent in those lower multicellulars, the sponges.

Smela (Kiev Region), July 1885.

REFERENCES

[1] Metchnikoff, E.I. *Issledovaniya o razvitii meduz i sifonofor.* Sobraniye sochinenij, Moscow: Izd-vo Academii Nauk SSSR, vol. II, 1946, str. 424–471.

[2] *Jenaische Zeitschr.*, Bd. VII, 1873, S. 471.

[3] Nabludeniya nad razvitiem kishechnopolostnyh. *Izvestiya O-va lubitelej estestvoznaniya, antropo* logii i etnografii, t. X, v. 2, M., 1874, str. 10.

[4] *Studien zur Gaestraea Theorie*, 1877, S. 267. Note.

[5] Embrional'noye razvitiye i pochkovaniye Pedicelina echinata, *Zeitschr. f. wiss. Zool.*, Bd. XXX, S. 525.

[6] Die Coelomtheorie, *Jenaische Zeischr.*, Bd. XV, H. 1, S. 124.

[7] *Issledovaniya o razvitii meduz i sifonofor.* (vol. 2 of Russian edition, pp. 424–471)

[8] These drawings are utterly similar to drawings in the Strassburger's book, *Zellbildung und* Zelltheilung, 3rd ed., Jena, 1880, Fig. II, Dr. 38–53.

[9] *Zeitschr. f. wiss. Zool.*, Bd. XXXII, 1879.

[10] Thus, I consider stages in Dr. 34 and 35 (Fig. XIX) to be too different to compare to each other; it is quite doubtful also that the two-layered endoderma becomes a one-layered one.

[11] *Handb. d. vergl. Embryol.*, Bd. I, Jena, 1880, S. 148, Note.

[12] Loc. cit., p. 338.

[13] *Das System der Medusen* (first part of the first chapter), Jena, 1879, S. 348. He mentions here a swiming-at-large gaestrula in Craspedota (p. 7, 119); since its existence is not proved by any data I will not consider it.

[14] O proishozhdenii kunin, pochkuyuscihsya v zheludke gerionid. *Izvestiya Ob-va lubiteley* estestvoznaniya, antropologii i etnografii, V. XXIV, M., 1876, str. 7.

[15] My observations of Cunina rhododactyla do nor confirm Haeckel's assumption that there is a genetic link between Carmarina and Cunina, as well as that formation of buds on Carmarina's tongue is due to sexless reproduction occured on the aboral surface of "maternal" bud. (This research is published in Vol. 2 of the Russian edition, pp. 424–471)

[16] Metchnikoff refers here to his *Spongiological Studies* published in Vol. II of the Russian edition, pp. 61–88.

[17] *Handb. d. vergl. Embryol.*, Bd. I, S.172.

[18] *Annales des sciences naturelles*, 3 Serie, t. VII, tabl. V, fig. 13–15, p.91.

[19] *Beitrag zur Entwicklungsgeschichte der Seeigellarven,* Heidelberg, 1849, S. 18.

[20] *Beobachtungen uber Anatomie u. Entwickl. einiger wirbellosen Seethiere*, 1851, S. 81,89.

[21] *Archiv f. Anatomie u. Physiol.*, 1851, Taf. XIV, Fig. 2,3, S. 347.

[22] *Archiv f. Anatomie u. Physiol.*, 1855, Taf. XX, Fig. 1–5, S. 493.

[23] *Annals and Magazine of Natural History*, v. XI, 1853, p. 122.

[24] Entwicklungsgeschichte der einfachen Asciden, *Mem. de l'Acad. de S.-Petersb.*, t. X, N 15, 1866, p. 5.

[25] Entwicklungsgeschichte der eingachen Asciden, *Mem. de l'Acad. de S.-Petersb.*, t.XI, N14, 1867, p. 5.

[26] Anatomiya i istiriya razvitiya Phoronis. *Zapiski Peterburgskoj Akademii nauk.* Prilozheniye N 1, 1867, p. 24.

[27] Beitrage zur Entwicklungsgeschichte der Holothuries, *Mem. de l'Acad. de Sci de S.-Petersburg*, Bd. XI, N 6, 1867, p. 4.

[28] Embryologische Studien an Wurmern und Arthropoden, *Mem. de l'Acad. de Sci. de S.-Petersburg*, t. XVI, N12, 1871, p. 28.

[29] *Goettinger Nachrichten*, N 7, 1868, p. 155.

[30] *Izvestiya Obschestva lubitelej estestviznaniya, antropologii i etnografii*, M., 1873.

[31] *Biologische Studien.* II Studien zur Castraeatheorie, Jena, 1877.

[32] *Handb. d. vergl. Embryol. Deutsch. von Vetter*, Bd. II, 1881, S. 306.

[33] *Entwicklungsgeschichte Beitrage*, Zeitschr. f. wiss. Zool., Bd. XXIX, 1877, S.231.

[34] Beitrage zur Entwicklungsgeschichte u. Morphologie der Anneliden, *Sitzungsber d.k. Akad. d.* Wissenschf., Bd. LXXIV, 1877, S. 446.

[35] Ueber Entwicklungsgeschichte von Teredo, *Arbeiten aus dem Zool. Institut in Wien*, Bd. III, 1880, S.26.
[36] Embryonalentwicklung u. Knospung der Oedicelina echinata, *Zeitschr. f. wiss. Zool.*, Bd. XXIX, 1877, S. 526.
[37] Studien uber Entwinclunggeschichte der Anneliden, *Arbeiten des. Zool. Institut in Wien*, Bd. I, H.3, 1878, S. 4.
[38] Studien uber Entwicklung des Ampioxus, *Arbeiten des Zool. Institut in Wien*, Bd. IV, 1881, S. 28.
[39] Ueber die Entwicklung der Tellerschecke, *Morphol.*, Jahrb., Bd. V, 1879.
[40] Die Coelomtheorie, *Janaische Zeitschr.*, Bd. XV, 1881, S. 1.
[41] Keimblatter u. Organanlage der Echiniden, *Zeitschr. f. wiss. Zool.*, Bd. XXXIII, 1879, S. 39.
[42] See: *Zool. Anzeiger*, 1881, N 78, S. 141.
[43] Die Coelomtheorie, *Jenaische Zeitschr.*, Bd. XIV, 1881, S. 120.
[44] Ibid., S.2.
[45] See: Metchnikoff, E.I. *Spongiologicheskiye issledovaniya.* Sobraniye sochinenij, Moscow, 1946, vol. III, pp. 61, 88.
[46] *Die Coelomtheorie*, S.78.
[47] *Handb. d. vergl. Embryol.*, Bd. II, S. 322.
[48] This drawing, as all the others, is sketched with a help of prisma; my major concern here, as everywhere else in drawing, is to depict contours in a most complete and accurate manner.
[49] *Beitrage zur Entwicklungsgeschichte u. Morphologie d. Anneliden*, Weiner Sitzungsber., 1877, S. 448.
[50] On embryology of *Polygordius flavocapitatus* see Repiahov's note in *Zool. Anzeiger*, 1881, N 94, S. 518. A conclusion that this annelid does not undergo any metamorphosis and that from a stage of worm it transforms directly into a stage of gastraea is based on misunderstanding.
[51] Note sur l'embryologie de la Salmacina Dysteri. *Comptes rendus de l'Acad. des sciences.* 17 janvier 1875, p. 2669.
[52] *Ibid.*, p. 518.
[53] *Anhandl. zur Entw. der Thieren*, H. 1, Leipzig, 1882, S. 86.
[54] Compare my studies on development of Echinodermata and Nemertea. *Mem. de l'Acad. des* sciences de S. Petersb., t. XIV, N 8, p. 49. A described there nemertin – Lineus Caoteus – is a species frequently encountered near Messina).
[55] *Studien zur Entwincklungsgeschichte d. Anneliden*, S. 104.
[56] *Handb. d. vergl. Embryol.*, Bd. II, S. 334.
[57] *Ibid.*, S. 343.
[58] *Archiv f. microse*, Bd. VII, 1871, S. 101.
[59] *Gastraeatheorie*, S. 20, 154.
[60] I intentionally ignore here mollusks, contradictory studies of which by Ray Lankester and Bütschli point to analysis of evolution of *Paludina.*
[61] Since I hope to publish data on creation of endoderma in intestinal-cavitals later, let me refer to the Russian-language work by Kovalevsky (1873) where he compares process of migration in *Obelia* (Eucope polystyla) to invagination in Acalefa, *Actinia*, and *Cerranthus.*
[62] *Handb. d. vergleichenden Embryol.*, Bd. II, S. 304.
[63] Nabludeniya nad razvitiem brachiopod, *Izvestiya Obscestva lubitelej estestvoznaniya, antropologii i* etnografii, 1874, str. 14,15.
[64] *Anatomie et developpement des Ophiures*, 1881, p. 72. If I reject delamination in *Ophiotrix*, then I should definitely agree with Apostolodes, who claims, contrary to my previous data, that embryo of *Amphiura squammata* has, together with anal hole, a rear intestine. I saw these formations after resuming my studies.
[65] Pro-hydra, discovered by Greev, was ignored by many authors. The same species was found last year by Reingard near Odessa, and was observed by me, though exceptionally rare.
[66] Notes on Embryology and Classification, *Quart. Journ. of Micr. Sc.*, 1877, p. 410.
[67] *Metagenesis und Hypogenesis von Aurelia aurita*, Jena, 1881, S. 12.
[68] I would like to mention here that, according to my studies, position of blastopore in this animal coincides with its future mouth.

[69] *Die Entwicklung der Keimblatter*, Festschrift, Leipzig, 1882, S. 38.
[70] Unters. uber Orthonectiden, *Zeitschrift. f. wiss. Zool.*, Bd. XXXY, 1881, S. 300.
[71] *Recherches sur l'embryologie des Bryozoires*, Lille, 187, tabl. III, p. 61.
[72] *Die Ctenophoren des Golfes von Neapel*, Leipzig, 1880, S. 106.
[73] On the development of ctenophores evolution see, "Nabludeniya nad pazvitiem kishechnopolostnyh", *Izvestiya Obschestva lubitelej estestvoznaniya, antropologii i etnografii.* M., 1873.
[74] *Mem. Acad. Sciences de St.-Petersburg*, t. X, N 4, 1866.
[75] *Nabludenuya nad razvitiem kishechnopolostnyh*, str. 36.
[76] *Ibid.*, str. 115.
[77] *Ibid.*, str. 197.
[78] Ueber den Bau der Ctenophoren. *Jenaische Zeitschr.*, Bd. XIV, 180, S. 330.
[79] Die Coelomtheorie, *Janaische Zeitschr.*, Bd. XV, 181, S. 121.
[80] *Ibid.*, S. 122.
[81] Usber die pathologische Bedeutung der intracellulaeren Verdauung, *Fortschritte der Medizin*, 1884, 17, S. 4.
[82] *Leçons sur l'histologie du systeme nerveux*, 1, 178.
[83] Ueber eine Brachiolaria des Kieler Hafens, *Arch. f. Naturgesch.*, 163, S. 242.
[84] Vertl. Entwicklungsgesch. d. Comatula medii, *Arch. f. mikrosc. Anatomie*, 1876, XII, S. 587.
[85] Issledovaniya razvitiya iglokoszih i nemertin.
[86] Zur Entw. der Holothurien, *Zeitschr. f. wiss. Zool.*, Bd. XXVII, 1876, S. 160.
[87] Stidiem ueber Entw. der Anneliden, *Arbeiten aus dem zool. Inst.*, Wien, 1, 1878, S. 95, 98.
[88] Keimblaetter u. Organanlage d. Echiniden, *Zeitschr. f. wiss. Zool.*, Bd. XXXIII, 1879, S. 45.
[89] Ueber Entwicklungsgesch. v. Teredo, *Arb. aus. d. zool. Inst.*, Wien III, 1880, S. 30.
[90] Comparative Embryological Studies, Part III.
[91] *Studien ueber Entwicklungsgeschichte der Thiere.* H. 2. Die Keimblaetter der Echinodermen, Wiesbaden, 1883.
[92] *Ibid.*, S. 59.
[93] *Zool. Anzeiger*, Bd. VII, 1884, S. 33.
[94] *Ibid.*, S. 100.
[95] On this occasion I would like to turn attention of my colleagues to the fact that a sea fauna in Tangere is quite poor by all means; during my 10-week stay there (from Decenber 1884 to February 1885) I caught from pelagic animals only one *Beroe*, several *Pelagia*, one completely degenerated kornerot and several *Craspedota* and *Euphausia*; as for a coastal fauna which can be studied only in time of extremely low tides it is quite poor too. As one more complication I have to mention an absence of experienced and reliable fishermen who could supply specimens difficult to obtain by researchers themselves.
[96] *Recherches sur la fecondation*, 1879, p. 202.
[97] *Embryology of the Starfish*, 1864. *Northamerican Starfishes*, 1877, Tab. 1, fig 24–28.
[98] I have to mention in passing that the part played by these structures in the formation of the walls of the body cavity was not outlined by Al. Agassiz, as the Hertwig brothers claimed (*Coelomtheorie*, S. 127). This renowned American scientist mistook the described embryonic sacs for the water-vascular system, while I found, in 1868, that they form the walls of the cavity; in the same year, I argued for their morphological parallel with the gastrovascular apparatus (*Bulletin de l'Acad. de St.*-Petersbourgh, 1968, XII, pp. 285, 298). I showed my preparations to my old friend A.Kovalevsky, with whom I was staying at that time in Messina, and he soon managed to extend this discovery to Sagitta. Since the Hertwig brothers' mistake was adopted in other works (*e.g.* in Wadeyer's paper in *Deutsche med. Wochenschr.*, N. 18, 1883), I have to draw attention to it.
[99] Issledovaniya o razvitii meduz i sifonofor.
[100] Ueber die Larve u. die Matamorphose der Echinodermen, *Vierte Abhandlung*, 1852, S. 31.
[101] *Studien ueber die Entwicklungsgeschichte der Thiere.* II Die Keimblaetter der Echinodermen, Wiesbaden, 1883, S. 44.
[102] Usually stages of development are marked according to the time passed from the moment of their artificial fertilization. In the meantime, the latter depends more on external consditions. Thus, for instance, creation of wandering cells in *Strong. lividus* in December, when temperature in my

Tanger's laboratory (located in an attic) in December was often about 40, occured only thirty hours after fertilization, while in Villefranca in April, the same result was achieved in the same species in just fourteen hours.

[103] *Zool. Anzeiger*, Bd. VII, 1884, S. 46.

[104] *Ibid.*

NOTES

1 More correctly Geryoniidae (based on the genus *Geryonia*). The Geryonidae is a family of crabs (based on the genus *Geryon*).

2 We have used "cleavage cavity" for "segmental cavity" throughout to avoid confusion with metameric segmentation (more relevant if the Geryonidae had been crustaceans). Both terms occur in 20th century works.

3 See M.T.Ghiselin and C.Groeben "Elias Metschnikoff, Anton Dohrn and the Metazoan Common Ancestor". *J. Hist. Biol.* 30:211–228, 1997.

4 Probably uric acid crystals.

5 Probably statocysts.

6 Williamson regards this as a case of larval transfer from a distantly related (bilaterally symmetrical) taxon.

7 Echinodermata: Holothuromorpha (larva).

8 Cnidaria: Scyphozoa.

9 Larva of *Phoronis.*

10 Echinodermata: Holothuromorpha.

11 Oligochaeta.

12 "Rusconi's anus = blastopore".

13 For a full account of Metchnikoff's attack on Haeckel's gastrea theory, se AI Tauber & L. Chernyak, *Metchnikoff and the Origins of Immunology* (Oxford, 1991), pp. 53–66; 89–90.

14 A genus of Bryozoa.

15 Probably for "flat".

16 Metchnikoff probably meant "after initial cleavage". In echinoderms, the first cell division is often 2 hours after the sperm enters the egg. It's a matter of definition of fertilization.

17 The object in question is the blastocoel.

18 In which the coelom is formed from splits in the mesenchhyme.

19 For "Annelida".

20 For "turbellarias". (The Turbellaria are unsegmented worms, but do not include the Nemertea).

21 For "Nemertina". "Nemertina" is one of the many alternative names for this phylum, but it is also a genus in the phylum, referred to later. "Nemertea" is less confusing.

22 "m" refers to mesoderm in the drawings.

23 Probably a prismatic drawing arm or *camera lucida.*

24 = medusoid forms.

25 *i.e.* it is not a reversion to an ancestral condition.

26 "Annelida" is here used to include not only segmented worms (*Polygordius*) but also unsgmented worms (*Lineus*). "Vermes" is a better term for this loose grouping.

27 Ctenophora.

28 Phylum Nemertea.

29 A blastula possessing both ectoderm and endoderm.

30 for "pectinates", which could be translated as "comb-jellies", but "pectinates" might be confused with mollusks (*Pecten*, Pectinidae). In this case, a Greek comb seems preferable to a Latin one.

31 There is no blastula and, therefore, no true blastopore.

32 ? for "moist camera".

33 or starfish and sea-urchins.

34 Most starfish have bipinnaria larvae, and in many this is followed by a second larva, the brachiolaria. These were originally generic names. We have followed the modern practice of not treating them as genera, but Metchnikoff probably did.

35 Holothuromorpha (sea-cucumbers).

36 Ophiuromorpha (brittle-stars).

37 = *Marthasterias glacialis*

[38] Brachiolarias succeed bipinnarias in the development of some starfish (but not *Astrpecten*). It is unclear which phenomena Metchnikoff had in mind.
[39] Larvae of holothurians (sea-cucumbers) which resemble the bipinnarias of starfish.
[40] Trachymedusae.

CHAPTER 6

EMBRYOLOGICAL STUDIES ON MEDUSAE[1]: ON THE EVOLUTIONARY ORIGINS OF THE PRIMARY TISSUES (1886)

PREFACE

Ever since I completed my studies on the Spongillidae[2] in 1876, my aim has been to develop a clear understanding of the successive stages in the development of the digestive system. Because I could not accept the "gastraea" theory, [I felt that] a new hypothesis on the origin of multicellular organisms was necessary. This was the objective of my studies on sponges, summarized in my *Spongiologische Studien* (1878), which served as guidelines for further research on lower animals and on intracellular digestion.[3]

Some physiological data from my program on medusae has already been published. Now, however, I intend to present the embryological part of my studies, which seem particularly important in relation to the origin of the endoderm.

Part of this research was conducted in the Zoological Centers of Naples and Trieste, and I am pleased to express my gratitude to their Directors, Professor A. Dohrn and Professor K. Klaus, as well as to their Inspector, Mr. Greffe.[4] I am also most grateful to Dr. J. Barrua, Director of the Zoological Station at Villefranche, for his generous help. Finally, my sincere appreciation goes to Mr. A. Helder for his efforts in preparing this work for publication.

Odessa, 11–23 February 1886.

Table of Contents

INTRODUCTION

Although it is not my primary purpose to write a monograph on the development of medusae, I consider it useful to start with a short historical review of publications in this field.

Early studies on the embryology of medusae were centered around a theory on the development of hydroid polyps, on which I feel impelled me to make some preliminary remarks.

The first scientific data on the development of hydroid polyps became available a century ago, when Kavolini [1] observed and described the "eggs" of a number of polyps. He noticed that these polyps grew within "ovaries", of a sort, and then crawled out of them. He also managed to observe further development of the "eggs", and claimed that this process was quite similar to the development of gemmae [buds]. Kavolini's first-class drawings allow us to conclude that the "ovaries" were partly gonangia, partly gonophores, and that almost all the forms which he called "eggs" were mature larvae. The only exception is the "eggs" of *Sertularia geniculata*, which, as is evident from his drawings (Figure VII, Drs. 3,4), consist of young *Obelia*[6] medusae. Kavolini had observed the development of medusae on a colony of hydroids. His interpretation of this process was, however, not correct: he simply reduced his observations to known phenomena in the development of lower animals.

The same applies to other 18th century zoologists, such as Pallas who described "eggs" of hydroids. His cursory observations refer to various organisms. His interpretations regarding a freshwater hydroid with strikingly conspicuous eggs are obviously not up to the accuracy of his observations and not correct.

It took a long time to piece together the elementary data on the life-history of medusae and hydroid polyps. Biologists in the late 18th century, such as Esper, Péron, Lesueur, Lamarck, etc., produced generalized systematic material and completely neglected the life-histories of the animals.

Consequently, when an important discovery was made in 1829, it was not immediately applied to developmental studies on medusae. Later, however, this discovery became a solid foundation for such studies. Michael Sars [2], Magister of Theology in Bergen, found, among other unknown marine animals, a polyp-like creature which he called *Scyphistoma filicorne*, together with another remarkable form, *Strobila octoradiata* Sars. This latter was considered by Sars to be a link between zoophytes and medusae, because of the small medusa-like swimming forms which detached from its polyp-like body. Later, Sars [3] came to the conclusion that *Scyphistoma* is nothing less than an early stage of *Strobila*, which gives rise to delicate medusae [ephyrae], and these latter are probably a developing form of some jellyfish. Soon after, Sars [4] confirmed his hypothesis that the forms produced by *Strobila* develop into *Medusa aurita*.[7]

Before this discovery was published, however, Ehrenberg and Siebold attempted to study the development of medusae experimentally.

Ehrenberg [5] located the ovaries of *Medusa aurita* and the eggs, the latter, as he observed, in pockets within fist-shaped pits and consisting of three[8] quite distinct forms. Some eggs looked like "blackberries", while others were "small, thick, mat-purple disks, resembling tiny medusae without tentacles or digestive channels". At that time Ehrenberg could not distinguish the sexes, and he assumed that the separate microscopic forms in the pockets [represented the two sexes] and that only the females could mature into adult jellyfish.

Soon after, however, Siebold described the real situation. He found male sexual products and deduced the process by which eggs from the pockets divided and developed into infusoria-like embryos. "The division of yolk described above is fully comparable to the division of the frog's egg, first described by Prevost and Dumas. Similar division occurs in nematode eggs prior to the development of the embryo." This comparison became a landmark in comparative embryology, leading to the understanding of the developmental stages of various animals.

Siebold was the first to identify the ciliated embryos as larval medusae, and this clearly influenced his ideas on the classification of the embryos. Later [7] he observed that the embryos settle and assume a sedentary hydra-like state, similar to *Strobila capca* – something he considered impossible in his earlier writings.

Only gradually [8] did it become clear that the swimming larvae of *Medusa aurita* change first into *Strobila,* then into delicate medusoids, and finally assume the basic form of adult jellyfish. These results were confirmed by Sars' discovery of similar development in *Cyanea*, and also by independent research on *Scyphistoma* and *Strobila* (*Hydra tuba*) by Daliel.

The elucidation of the complex developmental processes in hydroids was even slower, largely because of the diversity within this group. In 1832 in Trieste, Wagner [9] observed a hydra-like polyp, which we can now identify as *Podocoryne,* with capsules containing "eggs" with "horns" on their free ends. "I found it highly remarkable that their movements resembled those of medusae. I therefore consider them to be small medusae arising from the egg-capsules of polyps." Lowen, who published his observations two years later, studied a form, which he called *Syncoryne sarsii,* with "egg" capsules identical to those of *Podocoryne*, and this made a significant contribution to further studies. He identified forms similar to medusae within the capsules of females and considered it quite possible that they could separate from the colony and live freely. In addition, he noted that *Strobila octoradiata* Sars developed in a comparable manner, thus giving credence to the view that the development of jellyfish and hydroid polyps was analogous.

In Lowen's book there is a marvelous description of development in a species of *Campanularia* (later called *Gonothyraea*) which he regarded as a metamorphosis. He observed eggs with embryonic vacuoles and pigment spots, and worm-like, ciliated larvae, which his predecessors had interpreted as motile eggs. Although Lowen described these larvae, he was primarily concerned with their way of life. He drew their two lamellae (see Figure IV, Figs.14–18) and showed them as two-layered. He was also amazingly precise in observing how the larva attaches to the substratum, how it changes form, and how the hydrocaulus is formed.

These publications by Lowen, together with the works of Siebold and especially Sars (who, in 1841, discovered similarities in the development of medusae and salps) laid the foundation for the theory of alternation of generations established by Swenstrup a year later. He found a new species of *Coryne* in Iceland, with buds similar to those of medusae. This species, according to Swenstrup, produced eggs only after releasing buds.

Having compared the reproduction and development of zoophytes with other invertebrates, particularly trematodes, he concluded that hydroid polyps serve to produce nutriment for the sexually mature medusae which are generated by asexual reproduction, i.e. budding. Recently it has been claimed that Swenstrup's theory merely records the facts without explaining them, but it was quite influential in embryology because of it enabled predictions to be made and it opened up new horizons of research.

This also explains Siebold's enthusiasm for Swenstrup's theory of alternation of generations [14]. The only significant gap in this theory, namely the generation of nutriment [by forms arising] from the eggs of medusae, had been filled by Swenstrup. His prediction was confirmed some three years later by Dujarden [15], who managed to trace, although quite superficially, the development of a hydropolyp (*Stauridium*) from the eggs of a medusa (*Cladonema*).

After the normal succession of stages in the development of medusae was unraveled, it was still unknown whether this applied to all species of the group. As a result, a significant amount of [relevant] data was produced. It was already known that many hydroids do not have a medusoid stage in their development, and it soon became clear that some medusae could emerge directly from eggs, without an intermediate, hydropolyp, feeding stage. The first observations of this phenomenon were by J. Müller [16]. In 1851 he discovered a free-swimming larva of *Aeginopsis mediterranea*, covered in cilia, which he regarded as an embryo developed directly from the egg. The following year [17] he even suggested that some hydromedusae have no alternation of generations, while some go through a polyp phase.

This conclusion was soon verified when Krone [18] obtained larvae of the jellyfish *Pelagia* directly from eggs, and these grew into large jellyfish. This observation was quite amazing because *Pelagia noctiluca* is [systematically] quite close to *Chrysaora* in which alternation of generations had been studied by Daliel [19] and Bush [20].

It thus became clear that the processes of reproduction and development in medusae are extremely diverse and changeable. The explanation [for the delay in reaching this conclusion] was that research had been directed to systematic zoology for so long and embryology was not considered.

A number of studies have recently been undertaken to obtain definitive information on the reproduction of medusae and their connections with the polypoid generation. The fundamental work of Gegenbauer [21] is clearly the best of these; we shall consider the others later. He tried to introduce some order to a wealth of factual data, and he paid special attention to the sexual organs of those hydroids that do not produce medusae. He traced all imaginable transitional forms between medusae and [the contents of] sexual capsules and regarded these latter as an immature phase comparable with medusae.

In practice, Gegenbauer enriched our knowledge (a) by the study of a larva called by him *Trachynema ciliatum*, (b) by the discovery of the peculiar method of reproduction in *Cunina prolifera*, and (c), of special interest to us, by the study of the embryonic development of two species of Oceanidae, under the names *Lizzia Kollikeri* Geg. and *Oceania armata* Koll. From eggs of these medusae, he cultured infusoria-like, ciliated larvae (called "planulae" by Daliel), which attached to the substratum and changed into colonies of hydroid polyps. It is quite remarkable that his account of embryonic development contains hardly any references to embryological issues. (The only exception is that he considered the embryonic coelom and the formation of endoderm through [cell] division.) The description of the structure and orientation of the larvae by Gegenbayer was far less comprehensive than that by Lowen twenty years earlier. This neglect of comparative embryology can be explained in part by the preoccupation with zoology, and in part by the general stagnation which characterized embryology at this time. Even such important generalizations as that noted by Huxley of the similarity between the two layers of [diploblastic] metazoans and the embryonic layers of higher animals, and the formation of the enteron in *Pelagia* by invagination of the blastoderm were disregarded.

During the fifties and early sixties, a series of important studies filled many gaps in the understanding of reproduction in medusae and hydroids. These studies were, however, purely zoological. This refers to the numerous works of Allman, Hincks, Strathill Right, Al. Agassiz, Krone, J. van Beneden, Haeckel [22] and others. I shall return to some of these works later. [.....]

By the late sixties, [the application of] the theory of embryonic layers had spread to various invertebrates, thus providing a much broader foundation for the phylogeny of lower animals, so that comparative (i.e. interspecific) embryological studies finally became possible. Only then was the embryological analysis of the Metazoa initiated.

The first example of this is found in a short paper by Kovalevsky (1868) [23], in which he studied the major layers in some metazoans from a comparative viewpoint. He studied actinians and *Agalma* and also two medusae, *Pelagia noctiluca* and *Eucope*. Having described the formation of endoderm by invagination, he concluded: "This gives a rationale to comparing this method of forming the digestive cavity with that of the formation of the intestines of other animals, and the space between the wall of the enteron and the outer covering in *Eucope* should be considered as the body cavity of this medusa."

Kovalevsky's conclusion is quite consistent with the opinion of the talented but, unfortunately, prematurely deceased Russian zoologist, Nozhin, who, in his only published work [24], wrote that the formation of the endoderm on *Eucope* is so different that it "could hardly be interpreted morphologically[9] at all."

Kovalevsky did not publish a detailed account of his work until 1873 [25]. This book, in Russian, contains the following chapters on medusae: "Development of *Campanularia* from *Eucope polystyla* Geg.", "History of development of *Cassiopea borbonica* d.Ch., *Rhizostoma* Cuvieri and *Aurelia aurita*", and "On the history of development of *Pelagia noctiluca*". [...]

The ovum of *Eucope polystyla* (*Obelia polystyla* Haeck.) undergoes repeated segmentation and changes into a one-layered ball of cells, from which the formation of the endoderm is initiated. [...] After that, the embryo grows and becomes covered in cilia, while yolk cells accumulate at the rear end of the larva and gradually come to fill the entire cavity. In normal larvae the endoderm continues to grow and forms a large digestive cavity within the inner wall of structured cells. [...]

My studies on the Metazoa started before the publication of Kovalevsky's book. After studying development in the Echinodermata, I became convinced that there was a homology between the gastrovascular system [of coelenterates] on the one hand, and the peritoneal cavity, water-vascular system and alimentary system [of echinoderms] on the other. I did my research in the Rivira di Ponete (San Remo and Villefranche[10]), where I spent the winter of 1869/1870 investigating the development of some medusae (*Geryonia proboscidalis, Carmarina hastata* Hack., *Aeginopsis mediterranea* J.Müll., *Polyxenia leucostyla*) and several Oceanidae and Campanularidae.[11] The major results – separation of blastoderm into two embryonic layers in *Geryonia*, direct development of this medusa from the egg (also *Aeginopsis* and *Polyxenia*), formation of a mullberry-shaped embryo – were summarized in a short essay (1870). The publication of an enlarged volume (submitted to the Academy of Sciences in Saint Petersburg in 1872) was delayed until 1874, when it was published in *Zeitschrift für wissenschaftliche Zoologie* [26]. In the meantime, Folle published his outstanding study of the development of eggs in the Geryonidae, where he paid particular attention to the changes in the nucleus that occur during cleavage of the egg, in addition to establishing that both cellular layers are formed by division of the blastoderm. In his general review of the formation of embryonic layers in animals with an enteron, Folle came to the conclusion that they were formed by cell division [in a plane parallel to the surface of the blastula] (delamination) rather than by

Figure I

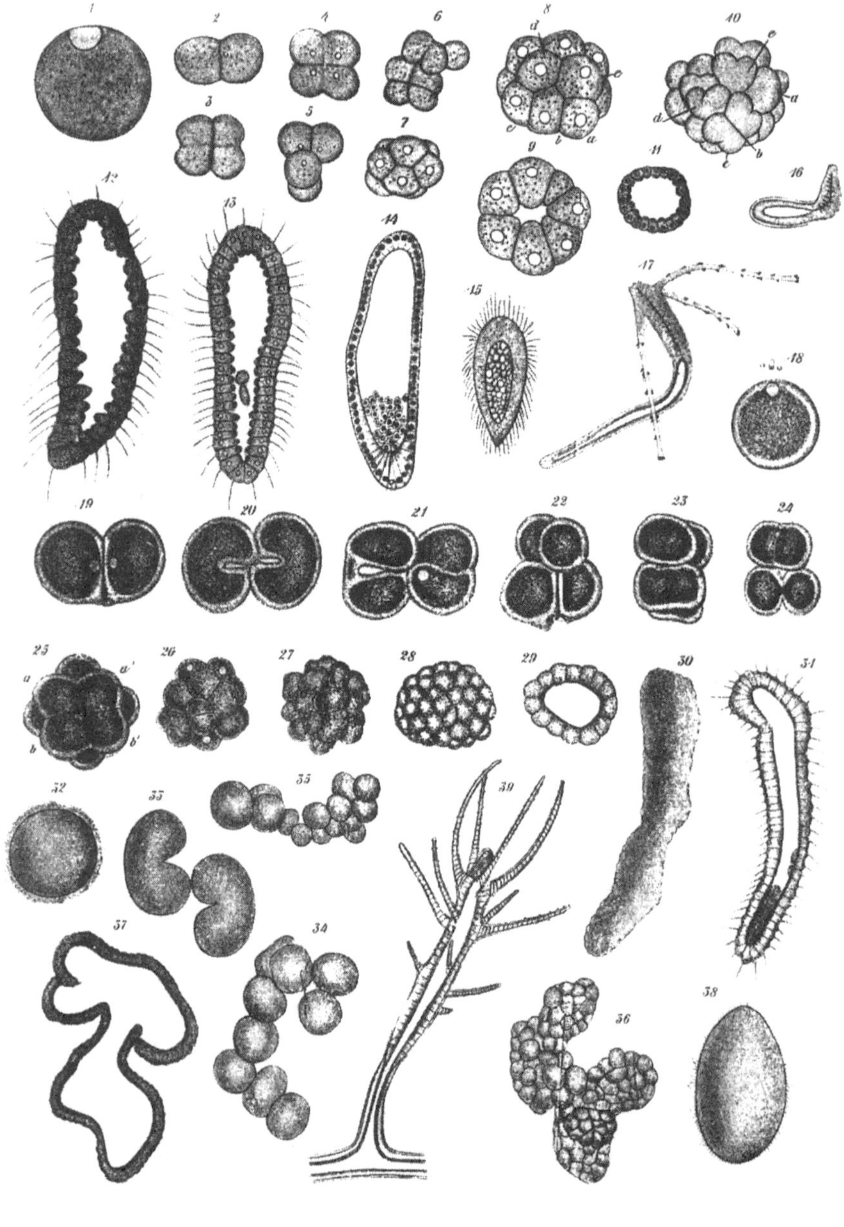

FIGURE I

Dr. 1–17 relate to *Tiara piteata. Al. Ag*

Dr. 7. 8-cell stage of division from the surface. 3+5.

Dr. 15. Two-layer planula. A live animal is depicted. 250:1.

Dr. 18–31 relate to *Rathkea fasciculata Häck.*

Dr. 20. Beginning of the second division of a newly laid egg. 200:1.

Dr. 21. The same egg in half an hour. 200:1.

Dr. 24. The same egg later. 200:1.

Dr. 30. Detached larva. 200:1.

Dr. 33. Beginning of the second division. Drawing is done without a drawing prism.

invagination.[12] An opposing view was put forward by Haeckel, who questioned this mode of [endoderm] formation. He claimed that those cases in which delamination has really been proved can be regarded as initial invagination modified by reduced inheritance or cytogenetic evolution. He considered it certain that endodem formation in many animals with an enteron occurs by invagination of the archiblastula, thus giving substantial confirmation to his gastraea theory.

The embryology of medusae has recently attracted the attention of a number of zoologists interested in the Acraspeda.[13] [.....] Some embryological data on the Acraspeda can be found in Haeckel's *Metagenesis und Hypogenesis von Aurelia aurita.* He claims that the different variations in development shown by this jellyfish include direct transition from gastrula to ephyra, without an intermediate scyphistoma stage. [....]

It is quite obvious from my review that the most important embryological processes in medusae show considerable diversity, and not enough is known to draw a general picture. How different, in this respect, are more complex animals such as, for example, ambulacrals[14] or mollusks, with similar embryonic processes of endoderm formation. It is clear that a closer look at medusae would extend our knowledge not only of their development but also of comparative embryology, and that is why I presented a comparative review of the major trends in this field in my introduction. I have made several expeditions to the Mediterranean in the last five years to gather information on medusae. I started my research in 1880 in Naples, where I studied a limited number of forms; from November 1882 to early May 1883, I worked at Messina, where I collected some crucial data and which proved to be a better place for the research presented in this book. To complete and verify my observations I went to Villefranche from February to May 1885.

This is the story behind the text, which, in spite of all my efforts, is still far from complete.

CHAPTER 1. THE EGG AND ITS FERTILIZATION

SHAPE, SIZE, TRANSPARENCY AND COLOR OF EGG – TIME OF LAYING – DEVELOPMENT OF EGG IN OVARY – MATURATION OF EGG – STRUCTURE OF LAID EGG – FERTILIZATION

Mature eggs of all species of medusae that I have studied are round, and *Geryonia proboscidalis* Esch. is no exception. I mention this species because Folle [27] describes its egg as oval. Some eggs can acquire an elongated oval shape under pressure or for other reasons, but usually eggs of *Geryonia* are completely spherical, as I mentioned in my previous work [28]. The same shape was observed by me in hydroids, with the exception of *Eudendrium,* which has peculiar pear-shaped eggs.

While the shape of the egg remains the same in almost all species, their size may be quite different, even in closely related types of medusae. An extreme case is presented by the Aeginidae. The largest egg in this family, up to 1.5 mm, is found in *Polyxenia albescens* (*Cunina albescens* Geg.), the smallest in *Cunina*. Between these two extremes fall all the medusae that I have studied. [....] It is quite apparent from this table that egg-size in medusae is not correlated with systematic position. [.....] The majority of medusae have transparent eggs – a characteristic that is obviously important for their protection. [....] While transparency is a regular feature in the eggs of medusae, it is a rare exception in those of hydropolyps. [.....] The specific gravity of the eggs of medusae is nearly always greater than sea water, so that they sink quite rapidly. [.....] The eggs of medusae are released into the sea by rupture of the ovary. The time of laying is specific in almost all types of medusae. [.....] The most favored time is from dawn to sunrise; eggs are not laid in the early morning, after sunrise. [.....] Interestingly enough, the time of egg-laying differs most in forms most closely related to each other. This can be explained, it is suggested, as an obstacle to cross-fertilization. [.....]

In the ovary, eggs originate as cells with a big spherical nucleus containing a small nucleolus. [.....] The eggs start to mature with changes in the nucleolus, which gives out a number of small particles (Figure IV, Drs. 17, 18). The next change is in the contents of the nucleus, where one can see chromatin granules and fine threads which become spindle-shaped (Figure IV, Dr. 20). [.....]

This leads to the formation of polar bodies, which may better be called polar capsules, which were first described in medusae by O. Hertwig.[15] []

After the formation of this spindle within the nucleus, an amoeboid outgrowth detaches from the nucleus (Figure IV, Dr. 26); it soon moves out of the egg, to which it remains attached by only a very fine thread (Dr. 22). [.....] When the formation of polar capsules is complete, medusan eggs are mature and ready for fertilization. [.....] Hertwig found that this "embryonic capsule" plays a significant part in directing cell formation, and its residuum can be identified within the nucleus of the egg. It was thus shown that the rudiment of the "first embryonic capsule" is present before fertilization and is independent of it. [.....]

Prior to fertilization of the egg, a lens-shaped furrow appears in the protoplasm nearest to the nucleus (Figure III, Dr. 22). Sperm streams into this furrow (Figure III, Drs. 22, 23) and promptly penetrates the egg; then, in live specimens, it is lost from view. Sometimes, on the other hand, it remains in the furrow for quite a while, and fertilization does not occur. [.....]

In unfertilized eggs that I have studied, there was either no sperm nucleus or it remained some distance from the egg nucleus. [.....] The theory recently developed by Hertwig [29] and others, that the nuclei of the egg and spermatozoon make up the idioplasm, whose theoretical existence was assumed by Negeli, receives no confirmation from studies on medusae. [.....]

All the above is important in understanding the current situation in the study of fertilization: notwithstanding the significance of recent results, they are still far from revealing the essential elements of fertilization and inheritance.

CHAPTER 2. THE FIRST THREE DIVISIONS

VARIABILITY OF THE INITIAL DIVISION – MERIDIONAL AND EQUATORIAL CLEAVAGE – SECOND MERIDIONAL DIVISION – CENTRIPETAL AND CENTRIFUGAL CLEAVAGE – THIRD DIVISION

It has been known for a long time that the eggs of medusae divide completely and more or less regularly to form a hollow ball – the blastula – or a solid morula. [.....]

Figure II

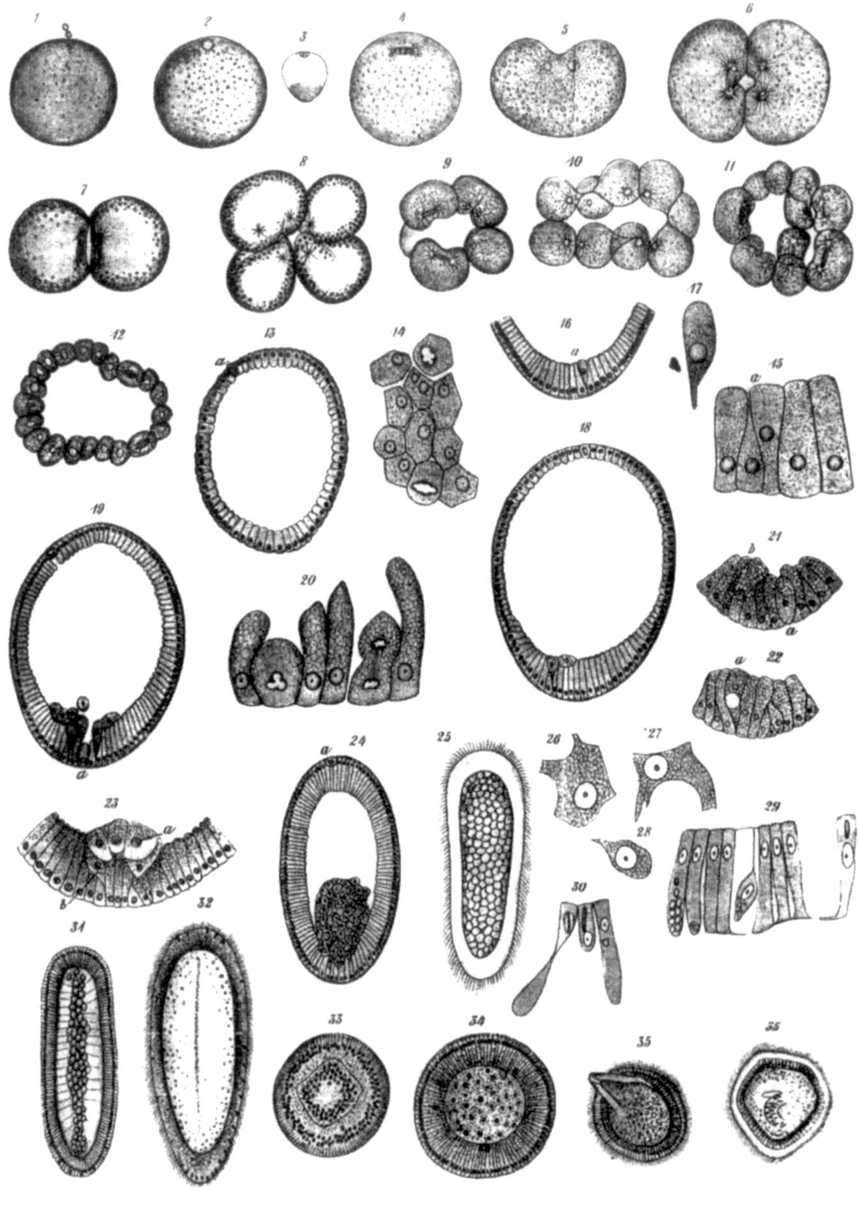

FIGURE II

Dr. 1–12, 21–23, 25–36 relate to *Clytia flavidula* mihi, Dr. 13–20, 24 – to *Clytia viridicans* mihi.

Dr. 4. Egg with dividing nucleus. Preparation is treated by vinegar acid. 105:1.

Dr. 9. Transition to the fourth stage of division in optic section. Preparation is treated by vinegar acid. 105:1.

Dr. 26–28. Three cells of endoderm of the same larva. 870:1.

Dr. 29–30. Cells of ectoderm of a more mature larva. 870:1.

Dr. 31. Larva on the second day after coming out of egg. Preparation is treated by osmic acid, Bil carmin, glycerin. 230:1.

Dr. 32. Larva on the third day after coming out of egg. A live animal is depicted. 115:1.

Dr. 33–34. Diametrical section across early larval stages. Preparation is treated by nicrosulphuric acid, borax carmin, and mounted in Canadian balsam. 390:1 and 350:1.

The embryology of medusae involves so many important problems that still await solution, and the division of the egg somehow became ignored and has only recently become the object of research interest. [.....]

The egg nucleus, as we have already seen, remains near the margin of the egg, where it comes into contact with a sperm nucleus. When fertilization is complete, the merged nuclei maintain their position near one pole (usually called the animal pole) where the polar capsules were developed. An equatorial nuclear spindle, with all the usual components (achromatic threads and a chromatic substance), is now formed, with star-shaped radiating bodies at both poles (Figure II, Dr. 4). Two new nuclei form and separate (remaining connected only by achromatic threads), and then the first furrow starts to appear in the ectoplasm. While the majority of scholars insist that the first cleavage furrow is meridional, and thus separates the egg into left and right halves, Haeckel [30], by contrast, contends that this fissure divides the egg into animal and vegetal cells and is, therefore, equatorial (as in some mollusks). The correct view is certainly the first, so a general rule may be established that the first cleavage is meridional in all known medusae. (Such a cleavage may also be called sagittal.) Cleavage in all medusan eggs that I have studied starts at the upper pole of the egg, apparently resulting from the nuclear spindle positioned next to it. [.....]

The timing of the first division is variable. [.....] The bigger the egg, the later the start of division. [.....] When the egg is completely divided into two equal blastomeres, with immobile capsulate nuclei clearly visible in live specimens, the cells remain in close contact. This spacing is quite obvious in hypogenetic medusae[16] with circular [holoblastic] sagittal cleavage (Figure VIII, Drs. 6–8), but less so in metagenetic medusae,[17] in which the blastomeres are not so clearly separated due to the superficial form of cleavage. If it is accepted that circular cleavage is more primitive than superficial cleavage, the origin of the latter form can be seen as a means of shortening development.

The second cleavage is also meridional, at right angles to the first; we shall call it a frontal cleavage. [.....] In *Rathkea fasciculata* (Figure I, Drs. 20,21), *Oceania armata* (Figure I, Dr. 33), *Clytia flavidula* (Figure II, Dr. 6), *Microtoma annae*,[18] *Laodice cruciata*, and *Nausithoe marginata* (Figure X, Dr. 6) the frontal cleavage originates between the blastomeres, next to the zone of contact between the two blastomeres. The cleavage plane is even more obvious because it penetrates the entire depth of the blastomere. Such a cleavage may be called centrifugal. [.....]

A study of frontal division supports the view that superficial cleavage evolved from circular cleavage, which is [the] primitive [method]. This is the original type for both centrifugal and centripetal cleavages, since the former cannot transform directly into the latter. [.....] In general, the deeper the cleavage that divides newly-formed blastomeres, the closer together are the neighboring (so-called "second cousin") blastomeres. Thus two pairs of blastomeres make up the four-cell stage (Figure VII, Drs. 10, 12 a+b, a'+b'); they move in pairs, and one pair often turns at right angles to the other. Since medusan eggs usually have no external membrane, a link between blastomeres is very important for their integration. Under such circumstances, the change from circular cleavage to superficial cleavage would give a tremendous advantage to a group of blastomeres. The significance of cellular integration for the survival of an animal is shown by examples of *Oceania armata* in which, by contrast, the blastomeres are in complete disorder. [.....] If the blastomeres are completely separated, no larvae are formed. Instead of embryos, only isolated blastomeres are found on the bottom of the vessel, and they soon die. Blastomeres of animals with an egg membrane do not need to be as close to each other; they retain the primitive circular method of cell division and still survive. [.....]

As already mentioned, dividing blastomeres show significant movement. [.....] Such movements, which may appear chaotic, are invariable. Perhaps they are important for molecular processes in cell division. [.....]

[The first] four blastomeres, which form a cross-shaped figure, undergo a third division, which is equatorial in all medusae. This division differs between different species, and does not necessarily follow the mode of previous cleavages. [.....]

When the equatorial division is completed, the blastomeres perform a peculiar move which results in the formation depicted in Figure I, Drs. 7, 24; Figure III, Dr. 29; Figure VIII, Dr.16; Figure IX, Dr. 8. They make up two cross-shaped figures, one of which lies on top of the other. These figures are not in line, so that the lower blastomeres can be seen in the spaces between the upper blastomeres if viewed from above or below. Further development results in more closely packed blastomeres, so this arrangement is no longer apparent. [.....]

This eight-blastomere stage is also interesting because some medusae develop a cavity [blastocoel] by splitting at this stage. Centripetal cleavage seems most likely to lead to the formation of such a cavity. [.....]

The first three divisions in hydroid polyps show significant similarities to the corresponding procedures in medusae. [....]

A number of medusae and hydroids from diverse environments have now been studied, and they all show a degree of self-differentiation (in the sense used by Roux [31]). Irrespective of the conditions in which the eggs develop (whether free in the sea or protected by gonophores, acrocysts, mucous membranes, or firm membranes) they all undergo the first three cleavages in the same morphological sequence, and differ only in some minor details. External factors, such as light and temperature, are of little significance. [.....]

Turning to the possible effects of gravity on the orientation of cleavage, my data clearly deny any direct involvement. [.....] I have observed that the first cleavage is not necessarily vertical in the eggs of a number of small medusae; it can be in any direction. If we call the first two cleavages vertical, it is not because they are literally vertical; they are so only with regard to the convention which defines the "upper" pole as that where the polar capsule is formed and where the sperm penetrates the egg cell. [.....]

Although, on the one hand, we see no significant effect of external factors on the first stages of embryonic development, there are, on the other hand, real differences between various forms. Centrifugal and centripetal cleavage, for example, are not determined by any external influence, and must be attributed to self-differentiation

In general, the early cleavage stages of medusae and hydroids, in spite of differences in their morphology and habitat, are basically identical, and therefore have their roots in basic similarities in the organization of their eggs. Amazingly, these

Figure III

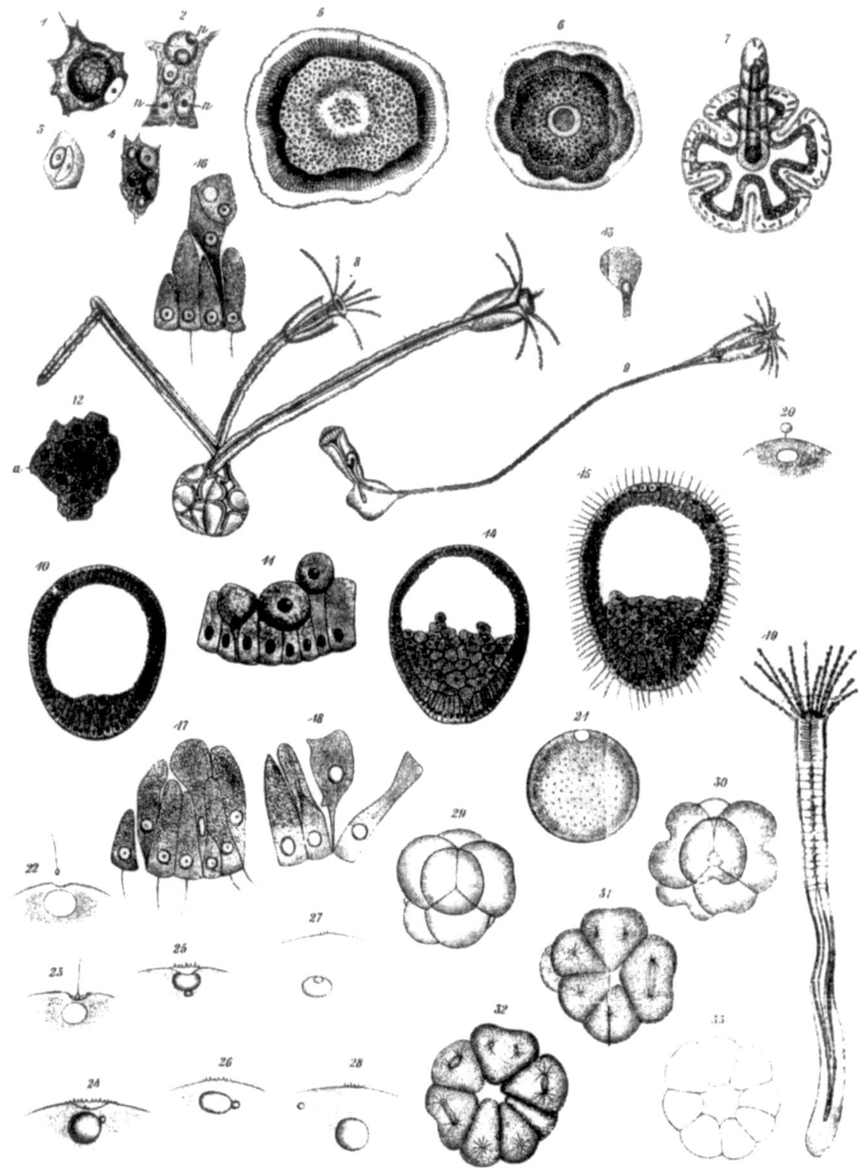

FIGURE III

Dr. 1–8 relate to *Clytia flavidula*; Dr. 20–33 relate to *Mitrocoma Annae.*

Dr. 2. Cell of endoderm depicted swallowing another cell. *N* – nuclei of swallowed cells. Preparation is treated with osmic acid, Bil carmin and glycerin. 870:1.

Dr. 3. Cell that creates a capsule. 870:1.

Dr. 20. Upper end of egg with a directing cell. 350:1.

Dr. 22. Upper end of egg with an approaching spermatozoid. 550:1

Dr. 23. Penetration of spermatozoid. 550:1. A live animal is depicted, Dr. 20–23.

Dr. 25. Stage with a sperm nucleus located below egg's membrane.550:1.

Dr. 26. Fertilized egg after leveling of depression. 550:1.

Dr. 29. 8-cell stage of division. 200:1.

Dr. 30. The same stage in an hour. 200:1.

Dr. 31. Transition stage between 3rd and 4th division. Preparation is treated by vinegar acid. 250:1.

forms of cleavage have been inherited by higher animals. Many animal forms, including the holoblastic eggs of vertebrates, retain the first three cleavage divisions typical of medusae.

CHAPTER 3. LATE STAGES OF DIVISION AND FORMATION OF ENDODERM

FOURTH DIVISION – BLASTULA STAGE AND FORMATION OF ENDODERM IN TUBULARIDAE AND CAMPANULARIDAE – EARLY DIFFERENTIATION IN *LAODICE* – VARIATION OF THIS PROCESS IN *POLYXEMIA LEUCOSTYLA* – GASTRULATION IN THE ACRASPEDA – FORMATION OF ENDODERM IN HYDROID POLYPS – TYPES AND SUBTYPES OF ENDODERM FORMATION IN DIFFERENT MEDUSAE

The fourth division, like the preceding ones, results from longitudinal cleavage of blastomeres, but shows considerable diversity. In the majority of cases, all eight blastomeres divide more or less simultaneously, but rarely is at as simple and regular as described by Klaus in *Aequorea Forskalii*. [.....] The majority of cleavage lines are partly meridional, partly equatorial, and sometimes in yet other planes (Figure II, Dr. 9; Figure III, Drs. 30, 31). Nuclear spindles, which give a quite reliable forecast of the cleavage plane, are placed tangentially, and usually towards the outside, although in some species they are closer to the central part of the blastomeres. They develop in a parallel fashion along either the longitudinal or equatorial axis of the egg, or they may be radial. [.....]

In *Aeginopsis mediterranea*, the eight blastomeres divide simultaneously but in different directions (Figure IX, Dr. 10). The sixteen cells resulting from this division are of basically the same size but with notable structural differences. A layer of ectoplasm, especially clearly marked in specimens fixed in osmic acid, is separate from the outer perimeter of each blastomere (Figure IX, Drs. 11–14); it includes very small granules, positioned regularly and orientated radially. A [further] layer of small but discrete granules can be seen at the border between ectoplasm and endoplasm. Further study of the sixteen-celled embryo shows that the ectoplasm is developed differently in different blastomeres: the more superficial blastomeres have a thicker ectoplasmic layer than those more deeply situated (Figure IX, Dr. 12). [.....]

In cases of unequal cleavage, i.e. when the upper blastomeres are significantly smaller than the lower ones, the former divide first (Figure VII, Dr. 18). [.....]

As we have seen, the sixteen-celled embryo is quite different in different medusae regarding the formation of the [blastocoel] cavity and the blastomeres: these latter may differ in relative size, positioning, and even structure. [.....]

Although it is quite clear that the formation of endoderm in all known metagenetic hydromedusae occurs through hypotropic migration (*hypotrope Auswanderung*) [= ingression], such a conclusion is not applicable to hypogenetic forms, as shown by my study of the embryology of *Geryonia*, *Polyxenia leucostyla*, *Aeginopsis mediterranea*, and from Folle's studies of *Geryonia*.

As already mentioned, the sixteen-celled stage in *Liriope mucronata* results from two meridional cleavages at right angles to each other. The diversity of forms that produce blastulas in this way leads to blastocoel cavities of quite significant size in some cases, while in others it may be almost filled by the ingression of blastomeres.

(Figure V, Drs. 12, 13). Close scrutiny of such embryos at this stage shows that already the cells differ in both shape and size. In *Liriope*, and also in *Geryonia proboscidalis*, the sixteen-celled blastula marks the end of this period of cleavage; the formation of endoderm starts soon after. [.....]

A cleavage cavity is still visible in *Liriope* at the start of endoderm formation, but it is soon filled by a growing number of endoderm elements (Dr. 17). [.....] I must emphasize that endoderm formation in *Liriope* is not localized in the embryo but takes place over the entire surface of the blastula,[19] as indicated in my previous works [32].

The embryo, in which the endoderm will subsequently be formed, is transformed into a dense parenchymella with two types of elements: ectodermal cells, formed of ectoplasm, and endodermal cells, formed of endoplasm. [.....] The parenchymella soon becomes a diblastula by the regrouping of the endoderm into a one-layered ball of cells (Figure V, Dr. 18). Although the history of development in hydroids resembles that of metagenetic medusae, endoderm formation in polyps follows the pattern of hypogenetic medusae, not the hypotropic migration of metagenetic medusae. [.....] While considering the well known fact that endoderm formation in hydroids is of the hypogenetic, not the metagenetic type, it should be mentioned that this process takes place at a stage when the embryos are motionless [.....], while in metagenetic medusae it occurs in the actively swimming stage. This latter circumstance [swimming] seems to limit endoderm formation to the rear end of the larva, while this type of development does not take place in the non-motile embryos of the former two groups.

The facts presented in this chapter indicate that endoderm formation in medusae is much more diverse than in any other group of animals. [.....] Only sponges show a variability close to that of medusae. In this group [the medusae], two major types of endoderm formation occur: all-round multipolar formation and hypotropic [polar] formation, localized at the rear end of the larva. The first type may manifest itself as either primary delamination, resulting from diametrical division of blastoderm cells (Geryonidae,[20] *Eudendrium*), or multipolar ingression (*Aeginopsis*), or secondary delamination, leading to the formation of a morula (*Aglaura*, *Rhopalonema*, many hydroids), or, finally, mixed delamination. Endoderm cells are often formed by diametrical division, or by subsequent differentiation, as in secondary delamination (*Polyxenia leucostyla*). This latter method of production represents a transition to epiboly, which leads us to a second type. Hypotropic production of endoderm is demonstrated primarily in the hypotropic ingession of metagenetic hydromedusae and in the invagination of many Acraspeda. Analysis of these types of development will be presented in the last chapter.

CHAPTER 4. LARVAE OF METAGENETIC MEDUSAE AND THEIR METAMORPHOSIS

CRASPEDOTA.: COMPOSITION OF LARVA – TWO TYPES OF METAMORPHOSIS: (1) POLYPOID FORMS AND THEIR SYSTEMATIC RELATIONSHIP; (2) LARVAE OF *NAUSITHOE* AND THEIR METAMORPHOSIS

The free-swimming larvae that originate from the eggs of all Craspedota with alternation of generations are genuine parenchymellas. The endoderm, formed by hypotropic

Figure IV

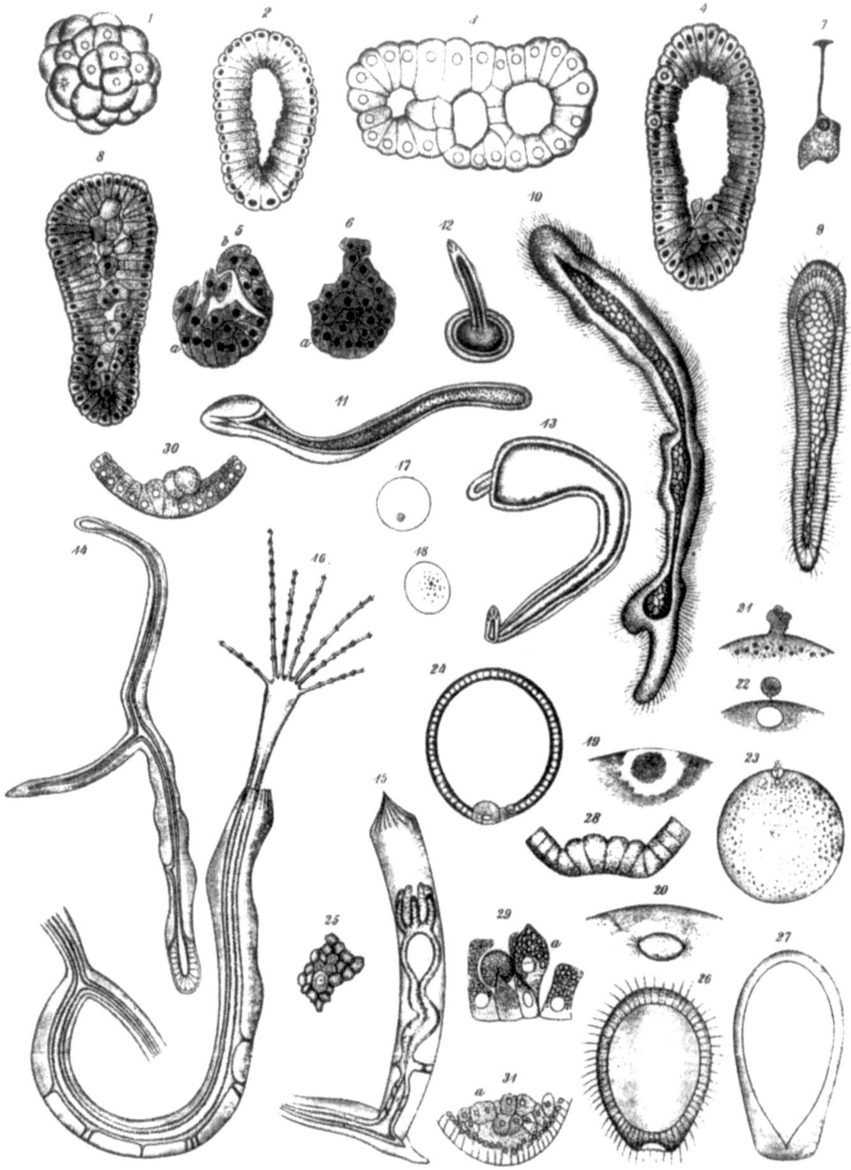

FIGURE IV

Dr. 1–16 relate to *Mitrocoma Annae Häck*; Dr. 17–31 relate to *Laodice cruciate L. ag.*

Dr. 8. Larva with almost filled cleavage cavity (cilia are nor depicted). Osmium fumes, Bil carmin, glycerin. 250:1.

Dr. 9. Swimming planula. A live animal is depicted.

Dr. 10. Large planula created by three blastulas.

Dr. 12. Another larva on 7th day of formation. 105:1.

Dr. 17. Embryonic bubble of egg from ovary. 160:1.

Dr. 18. Another bubble after dissociation of a small nucleus was initiated. 160:1.

Dr. 20. Upper end of the same egg from ovary after treatment by osmic acid. 550:1.

Dr. 22. Upper end of another egg after dissociation of a directing cell. A live animal is depicted.

Dr. 24. Stage of blastula at the beginning of swimming period. 170:1.

Dr. 26. Swimming blastula (18 hours of formation). Preparation is treated by osmic acid fumes. 170:1.

Dr. 27. Further development of larva after disappearance of the white pole. 170:11.

Dr. 28. Rear end of larva with white pole. Preparation is treated by vinegar acid, 230:1.

ingression, fills the entire cleavage cavity [blastocoel], and consists of polygonal or round cells connected to each other in an apparently random order. As for the ectoderm, it is represented by a single layer of cells with fine flagella evenly spread over the entire surface. The parenchymella stage is similar in all medusae that I have studied: it is ovo-cylindrical, wider at the front end and a narrower posteriorly (Figure I, Dr. 15 *Tiara*; Dr. 30 *Rathkea*; Figure III, Dr. 25 *Clytia flavidulum*; Figure IV, Drs. 9,10 *Mitrocoma*). Medusae with transparent eggs also have transparent larvae; by contrast, larvae of *Oceania armata, Rathkea fasciculata* and *Laodice* appear black or brown-black viewed against a light source. Evenly covered with long flagella, these larvae swim in spirals, with the animal pole always directed forward. Histological analysis of transverse sections (Figure II, Dr. 34) show the ectoderm cells in a thin layer, each looking very like a compressed parenchymella, and clearly divided from the endoderm cells with their rounded margins. [.....] The endoderm consists of relatively large parenchyma cells with various protoplasmic protrusions (Drs. 26–28). [.....]

While some larvae (e.g. *Microtoma Annae*, Oceanidae, and *Laodice cruciata*) cease their development at this stage, others (e.g. *Clytia*, *Obelia*, and *Aequorea*) undergo further changes. On the second day of their free-swimming period, the first stinging capsules [nematocysts] are developed at the posterior end (Figure II, Dr. 31). These cysts are formed within ectoderm cells from peculiar nucleus-like bodies near the true nucleus (Figure III, Dr. 3) and acquire their characteristic spindle shape with fixed outlines (Figure II, Drs. 29, 30). Some interesting developments are also to be seen in the endoderm. Peripheral cells with large granules group to form a regular layer, while cells nearer the center retain their original parenchymatous state (Figure II, Dr. 31). Examination of squashed larvae shows that endoderm cells fight with each other, which leads to the occasional swallowing of smaller cells by larger ones. Quite often endoderm cells already contain other cells, in addition to the one being swallowed (Figure III, Dr. 2), and these cells can easily be recognized by the color of their nuclei (nn in Dr. 2). Many endoderm cells are completely full of bodies of different size and origin, which should probably be regarded as remains of captured cells. There is thus intracellular digestion in the endoderm, like that which occurs at metamorphosis in the larval development of different animals. Inner cells are being devoured by larger outer cells, and, in place of the former, a narrow, fissure-life cavity can be seen (Figure II, Dr. 32). [.....]

In the larvae I studied, I did not see as big an endodermal cavity as that of the old larva of *Obelia* illustrated by Kovalesky (see his book, Figure I, Dr. 9). The dark granules on the outer layer of endoderm cells (also observed by Merezhkovsky) were already deposited at the time of formation of the cavity, and should be regarded as the remains of intracellular digestion (cf. Figure II, Dr. 32). [.....]

We thus see that a high degree of organization is inherent not only in free-swimming larvae but also in the subsequent sessile stage: the layered larva changes into an independent unit with obvious antimeres.[21] If a medusan larva could metamorphose into such a plate-like form, with all its muscles intact and retaining its free-swimming way of life, we would have a form corresponding to a medusoid developing from a hydrorhiza. Perhaps something like this happened to produce some siphonophores. [.....]

The hydroids that develop from the eggs of medusae belong to different groups [from the medusae].[22] Those that develop from eggs of *Tiara leucostyla* should be classified in the family Clavidae. [.....] The hydroid of *Oceania armata* also belongs to the Clavidae, [.....] All its characters are in agreement with such an affiliation, which is even more evident than in the case of *Tiara leucostyla.* [.....]

The hydroid *Tima formosa*, described by Al. Agassiz [33], seems to be closely related to *Campanopsis*, and differs from it principally by the less developed hydrotheca which extends only to the base of the polyp. In this respect it is closer to *Aequorea* than to *Octorchis*. Hydroids of *Tima pellucida* (*Eurene pellucida* Haeck.), cultured by me from eggs in 1870, showed marked similarity to the corresponding stage of *Aequorea* (i.e. the form with hydranths) in the development of a basal membrane of tentacles and the formation of a small peridium enclosing the base of the polyp. The only difference, although significant, is that the larva of *Tima* metamorphoses to a hydrorhiza, from which the hydranth buds.

Thus, my recent observations confirm the theory developed by my wife and myself [34] on the close affinity between the Geryonopsidae and the Aequoridae.[23] [.....]

Recently Klaus [35] expressed support for [the view that there is] a close relationship between the two families. I must emphasize that, when referring to the Aequoridae, I do not mean the entire family in the vast sense it acquired in Haeckel's system, but only *Aequorea* and related forms. According to Haeckel [36], the sole distinguishing character of the family Aequoridae is the large number of radial canals, a feature developed quite independently by many medusae. Using the number of radial canals as the main criterion for division, it is apparently impossible to separate the Melicertidae and Oschistomidae from the Thaumantidae as independent families, and, equally unreasonably, it would divide the Campanularidae (with diverse numbers of canals) into several families. I think that more careful study of the so-called Aequoridae, paying particular attention to the hydroid generation, will show that this family is not a unified entity, and that the majority of its members are related to the Geryonidae, others to the Thaumantidae, and yet others to the Eucopidae.[24] [.....] Closer scrutiny will, very possibly, remove *Zygocanna* and *Zygocannota* Haeck. from the artificial family Aequoridae. Strictly speaking, dissolution of the latter was initiated by Haeckel himself, when he ascribed *Orchistoma* (originally classified in the Aequoridae) to the Thaumantidae. [.....]

If a study of phylogeny compels us to exclude the Geryopsonidae from the Eucopidae (*sensu* Haeckel), the same approach forces us to remove *Mitrocoma* from this family, and also the related forms *Tiaropsis, Phialis* and *Halopsis*. Only forms related to the Campanularidae ((*sensu* Hincks) should be considered as genuine Eucopidae. The first of these forms are *Obelia* and *Clytia.* As L. & A. Agassiz [37] showed more than 20 years ago, many species of Gegenbauer's *Eucope* should be regarded as young medusae of *Clytia*, and the same opinion was expressed by Hincks. Haeckel [38] later claimed that one species of *Eucopea* (*E. thaumanthoides*) is the young form of *Phialidium* and it develops from [the hydroid] *Campanulina tenuis* v.Ben. A medusan [species of] *Clytia* was selected by him as representing a new genus *Eucopium.*[25] This proposal was recently opposed by Klaus [39], according to

Figure V

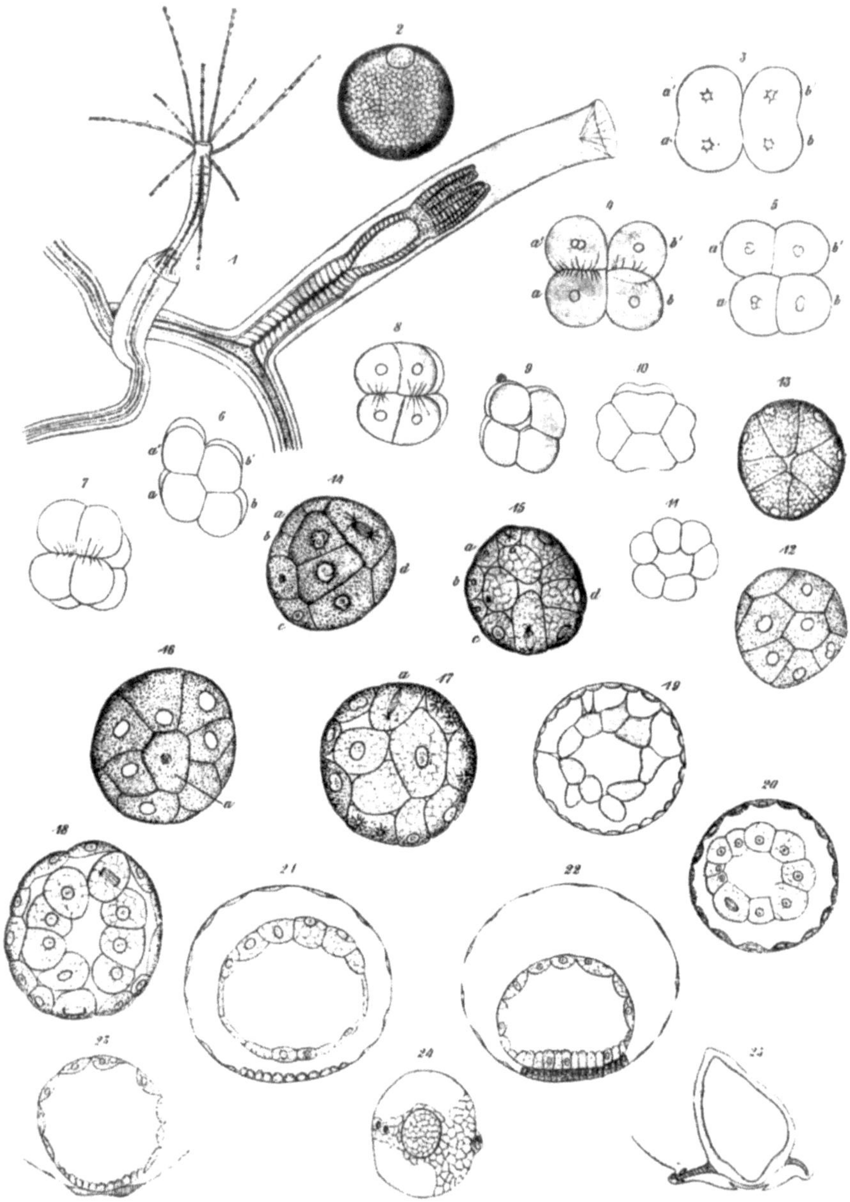

FIGURE V

Dr. 2–25 relate to *Liriope mucronata Gep.*

Dr. 12. 16-cell stage from the surface. Preparation is treated by vinegar acid.. 200:1.

Dr. 13. The same object in optic section. 200:1.

Dr. 17. Embryo in optic section. 280:1.

Dr. 18. Embryo with almost complete delamination. Preparation is treated by vinegar acid. 280:1.

whom *Phialidium* is descended from *Clytia*, not *Campanulina*, and, further, that *Eucopium* is not a separate genus but merely a variety of *Phialidium*.

My own observations on *Clytia flavidulum* and *viridescens* show that *Phialidium* is indeed not related to *Campanulina* and develops from genuine *Clytia*; therefore the generic name *Phialidium* should be changed to *Clytia*. [.....] Description of young, freshly released medusae of *C. flavidulum* are given in my "Medusologische Mittheilungen" (Figure I, Dr. 15), to which the reader is referred. [.....]

My observations on the metamorphosis of *Nausithoe marginata* are admittedly scant, but they are quite sufficient to conclude that this jellyfish has a succession of generations. The young scyphistoma excretes a cuticular membrane into which the animal can retract if necessary. Such specific features bring this animal close to the peculiar creatures described by Allman [40] as *Stephanoscyphus mirabilis* and by Schultze [41] as *Spongicola fistularis*. As pointed out by Schultze, this particular animal form corresponds to a scyphistoma, and was, in fact, recognized as such. [.....]

The fact that neither Allman nor Schultze found gonads in this animal, and also that *Stephanoscyphus* is distributed over the area where *Nausithoe* lives (the Mediterranean Sea) is consistent with my hypothesis that this form of scyphistoma is part of the developmental cycle of the above-named medusae.

CHAPTER 5. LARVAE OF HYPOGENETIC MEDUSAE GERYONIDAE – LARVAE OF *AGLAURA* AND THEIR METAMORPHOSIS TO THE SO-CALLED *TRACHYNEMA CILIATUM* GEG – SOME REMARKS ON LARVAE OF *PELAGIA NOCTILUCA*

While related forms in the Acraspeda (*Chrysaora* and *Pelagia*, for example) develop in entirely different ways, development in the Craspedota shows much closer links to systematic position. Thus, hypogenesis[26] is here limited to the trachymedusae and the Aeginidae (in the broad sense), which apparently have no alternation of generations. Haeckel's assumption [42], namely that such medusae (which he calls Trachylina) only occasionally undergo metagenesis,[27] is not confirmed by any data; four marginal cysts in the young medusa of *Lovenia clausa* Hincks cannot be regarded as sufficient reason to attribute it to the Trachylina. The other features of this medusa all agree with those of the Campanularidae [43]. Another of Haeckel's ideas to which I do not subscribe is his [44] insistence that *Ametrangia hemispherica* Allm., which is related to the Cannotidae Dipleurosoma, develops directly, without alternation of generations. Allman's account [45], to which Haeckel refers, shows that this medusa is viviparous and that its planulae differ from the larvae of typical hydroids by their oval body and absence of cilia[28] [? flagella] . Thus it appears (for now at least) that the assertion that species of the Tubularidae and Campanularidae (Leptolinae Haeck.) develop metagenetically, and trachymedusae and the Aeginidae, by contrast, develop hypogenetically, is correct, although it is evident *a priori* that there must be exceptions to this rule. [.....]

I consider Ray Lankester's attempt [45] to justify Haeckel's view as unsuccessful. The British researcher accuses Folle and myself of ignoring the stage with four tentacles drawn by Haeckel in Drs. 29 and 30, and jumping immediately from the stage without tentacles to the stage with six tentacles. He thus assumes that our knowledge is incomplete and that the lacuna is filled by Haeckel's data. This misunderstanding can easily be resolved: Haeckel's drawings relate to the tetraradial *Liriope*, while those of Folle and myself refer to the hexaradial *Geryonia*. Since the latter develops six tentacles directly, the tetraradial stage is completely absent. After I investigated *Liriope* (and, therefore, its four-tentacled larvae), which Haeckel also studied, I am even less inclined to accept his data. Besides, Haeckel does not claim (as Ray Lankester thinks) that the young stages of *Liriope* have closed subumbrellar cavities; my latest observations show that this claim (based on analogy with *Limnocodium*) is certainly incorrect. [.....]

Since larval stages of *Aglaura* are often captured by Muller's pelagic net, it is not surprising that they became the object of research interest long ago. J. Müller discovered them not far from Nice in 1849, and he then described and drew them. Although Müller did not classify these stages, he considered them to be young medusae. [.....]

Later [46] he suggested that these larvae were part of the developmental cycle of *Polyxenia leucostyla* Will. ; this, however, is not the case, since they should be regarded as young *Aglaura*. A few years later, the same larvae were described by Gagenbauer [47] as *Trachynema ciliatum*. Gagenbauer made use of stages discovered by his predecessor and also observed older *Aglaura,* with eight radial canals and numerous tentacles, but differing from adult individuals in having a two-lipped mouth and no manubrium. The presence of a small number of tentacles (three or four) in what Gagenbauer regarded as younger stages may be explained by the [assumed] loss of other tentacles, which frequently happens to net-caught larvae. After our first encounter with ciliated *Aglaura* larvae fifteen years ago, my late wife and I suggested that they were identical with *Trachynema* and that this latter name should be deleted from the system.[29] My recent research completely confirms this suggestion. I cite here, as a demonstration that different varieties of *Trachynema* have little in common, the attribution to *Trachynema* by Al. Agassiz of a form very close to *Aglaura* (*Trachynema digitale* = *Aglantha digitale* Haeck.), and, similarly, Haeckel's inclusion of *Smithea* Gegenb. in genuine *Trachynema*.

My recent observations on hypogenesis in the Aeginidae (*Polyxenia leucostyla* and *Aeginopsis mediterranea*) completely corroborate my earlier data, as well as the accumulated data of J.Müller [48]. The only fact that I missed then and have to add now [49] is that *Aeginopsis* also has a larval stage; this latter is stick-shaped (Figure IX, Dr. 27), as I observed in *Polyxenia* but not in *Aeginopsis.* [.....]

It was Krone [50] who, using *Pelagia noctiluca*, first demonstrated that some jellyfish develop directly from the eggs. This was later confirmed by Agassiz (using *P. cyanella*) [51], Kovalevsky (using *P. noctiluca*) [52], and Haeckel (using *P. perla*) [53]. My own observations are on *P. noctiluca*, whose eggs I obtained on many occasions but only once did I observe its normal development. I was particularly interested in confirming Kovalevsky's observations on the detachment of certain cells from the endoderm. At the time, I thought that this process produced a rudimentary mesoderm, but this proved not to be the case. I discovered that what I had taken for rudimentary mesoderm was merely a non-cellular residue formed during the

Figure VII

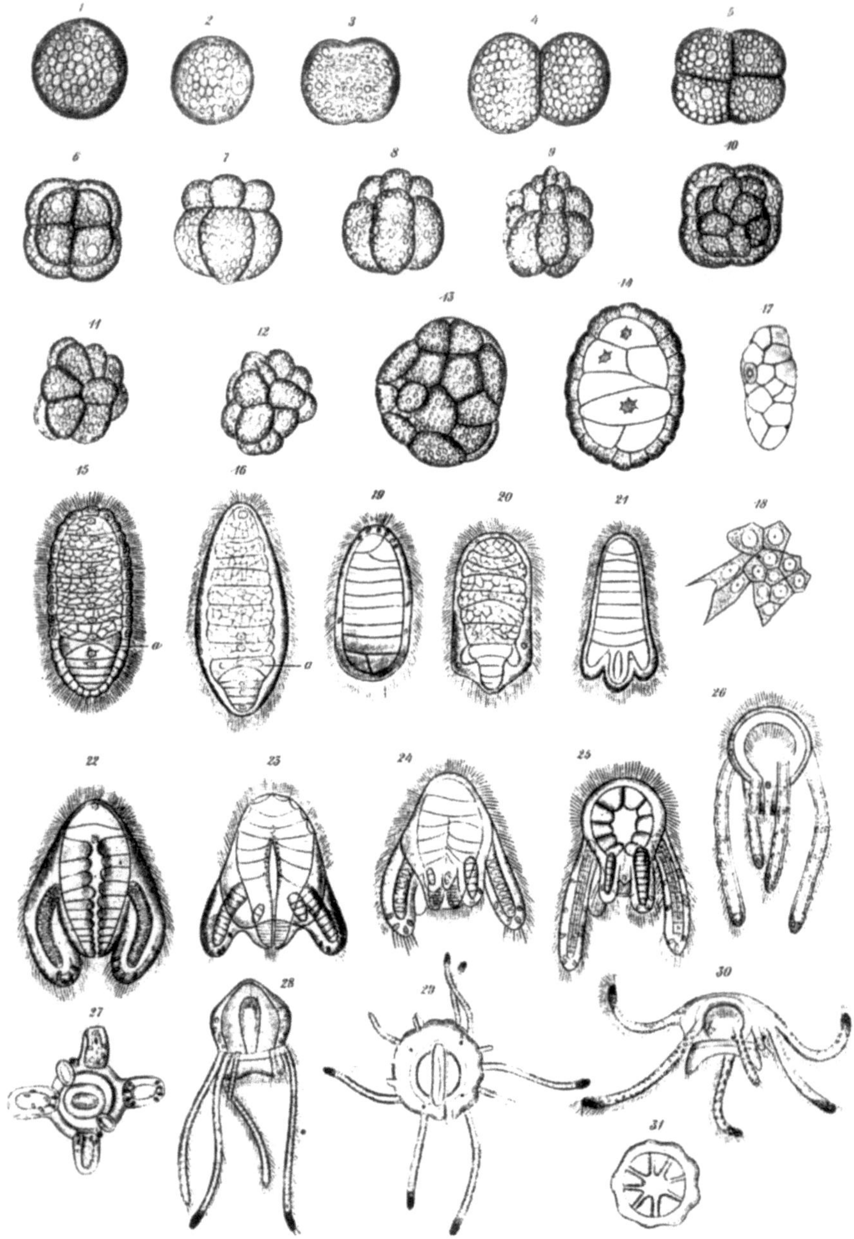
1
2
3
4
5
6
7
8
9
10
11
12
13
14
17
15
16
19
20
21
18
a
a
26
22
23
24
25
27
28
29
30
31

FIGURE VII

Dr. 1–31 relate to *Aglaura hemistoma.*

Dr. 10. Embryo at early stage. A live animal is depicted. 250:1.

Dr. 12. The same embryo in half an hour. A live animal is depicted. 250:1.

Dr. 18. Endoderm cells of the same larva, treated by osmic and vinegar acid. 540:1.

preservation of the larva (Figure X, Dr. 2). Gastrulas which I obtained were considerably elongated (as described be Krone), and were similar to the larval stages [that settled as] scyphistomas of *Cotolorhiza*. Longitudinal sections of *Pelagia* larvae (Dr. 24) show quite convincingly that there is no cellular material between ecto- and endoderm. All three (Krone, Agassiz and Haeckel) observed the simultaneous appearance of eight denticulate lobes (Dr. 25), accompanied by considerable widening and shortening of the larva (Dr. 26).

Although the involvement of endoderm in the formation of the lobes can be seen in a live specimen, longitudinal sections are more revealing (Dr. 27). The larva, by shortening of the body, turns into a completely flat disk, which is the beginning of the ephyra stage.

CHAPTER 6. ASEXUAL RERODUCTION AND BUDDING IN *CUNINA PROBOSCIDEA* METSCHN.

HISTORY OF "INNER" AND "OUTER" BUDDING IN *CUNINA* – ON THE SEXUAL PRODUCTS OF *C. PROBOSCIDEA* – EMERGENCE AND DIVISION OF AMOEBA-LIKE CELLS – ASEXUAL REPRODUCTION AND BUDDING – DIMORPHISM OF *C. PROBOSCIDEA* – RELATED FORMS – ANALOGY OF ASEXUAL REPRODUCTION IN *CUNINA* WITH THAT OF TREMATODES, AND WITH PEDOGENESIS IN *MIASTOR*

The species of *Cunina* are medusae whose embryonic development is almost completely unknown. Although I succeeded in obtaining artificial fertilization on a number of occasions, the embryonic development of *Cunina* is still mystifying because of small gaps that have never been filled. Asexual reproduction is much easier to observe, especially that form which became known more than 30 years ago as inner budding. [.....]

Important data were accumulated by Fritz Müller [54]. In the gastrovascular system of the majority of 8-radial *Cunina Kollikeri*, he found different stages of similar but 12-radial medusae, considered by him to be a by-product of budding from the wall of the stomach. In many cases, and this is a quite amazing fact, this peculiar type of reproduction was observed in mature male specimens with active sperm. This was considered as budding, and hence dimorphism, and this was assented to by myself. My [present] opinion [however], based on budding in the Aeginidae and presented most clearly by Haeckel [55], is that such a claim cannot be sustained. In the gastrovascular cavity of *Cunina rhododactyla* Haeck. (= *Eurystoma rubiginosum* Koll. [55]) I found numerous young stages, including the youngest with just one tentacle, loosely arranged in the mother's body. Unable to trace the origin of these embryos, I joined my predecessors in assuming a link between these spore-like embryos and the mother's body, and I considered this as a case of "inner" budding. Although I observed *Cunina* with young stages that had a significant number of sectors, I did not regard this as proof of dimorphism because of the considerable individual variation in such medusae. [.....]

Contemporary researchers, led by F.E. Schultze [57], consider as insufficient the arguments in favor of genuine budding, which apparently takes place in such cases, and they suggest that the observed embryos were developed from eggs. Haeckel [58] reiterates this position and emphasizes the possibility that "young cunanthid larvae developed from eggs [of one female] penetrate the gastric cavity of another cunanthid quite early, at the gastrula stage, and mature there as parasites".

Another type of reproduction frequently observed in *Cunina* is outer budding. [.....] Haeckel [59] studied this phenomenon and called it allogenesis. He regarded the formation of *Cunina* as taking place by budding on the manubrium of both sexually mature females and juveniles of *Carmarina* (*Geryonia*).[30] [.....]

After this brief historical sketch on asexual reproduction in *Cunina*, I consider it important to turn to sexual development in this medusa.

While earlier researchers, like Kölliker and Gegenbauer, gave no information on the sexual organs of *Cunina*, F.Müller [60] made original studies on mature *C. Kollikeri*; he, however, never observed eggs, only males with motile sperm. In *C. rubiginosa* (*C. rhododactyla*), Haeckel found one female specimen with eggs that were "big and distributed in small numbers at the edges of the radial gastric pouches, so that each pouch had 5–10 eggs on average." Males were also observed far more often by Haeckel [61]. I also had an opportunity to verify that the gonads are concentrated in the lower part of stomach. [.....]

The sexes are separate in both the species of *Cunina* (*proboscidea* and *rubiginosa*) that I studied with respect to their gonads. Each testis contains a countless mass of small cells, and each sperm has a quite large, rounded head, a small middle region, and a long flagellum (Figure XII, Dr. 10a). Mature sperm occur in shear-shaped packages, oriented in different directions, next to and above each other. In *C. proboscidea*, formation of the gonads takes place in the open spaces of the gastric pouches, spreading to the edges and forming numerous folds; they do not, however, enclose the so-called peroneal canals. [.....]

I conclude from all my observations on *Cunina proboscidea* that immature sex cells of many types move from the regions where they originate to penetrate the endoderm of the gastric pouches, peroneal canals and ring canal, as well as the sub-umbrellar jelly-like substance [mesogloea]. This process takes place in young males, adult males with developed sperm, and females, and is linked – and I want to emphasize this – to sporogonial reproduction. To understand the behavior of these migrants, the wandering cells of the mesogloea should be studied first: they are isolated and thus easy to observe. [.....]

Each time I had a chance to observe division in these wandering cells, they separated from each other to varying distances, but they never remained as close as in the first cleavage of egg cells. Fusion of cells, which I observed as well as their division, took place only in those wandering cells in which there was no nuclear activity and small granules and the nucleolus were clearly visible. [.....]

Although I studied fusion of wandering cells immediately after their division, I cannot prove that these two processes are intimately linked to each other. It is still possible, however, that touching cells, which acquire such significance later, are daughter cells with completely formed nuclei. Sporogony may very well be initiated by the division of wandering cells. It is well known that at the beginning of this process one cell absorbs another. The ingesting cell serves as a protective layer and

Figure VIII

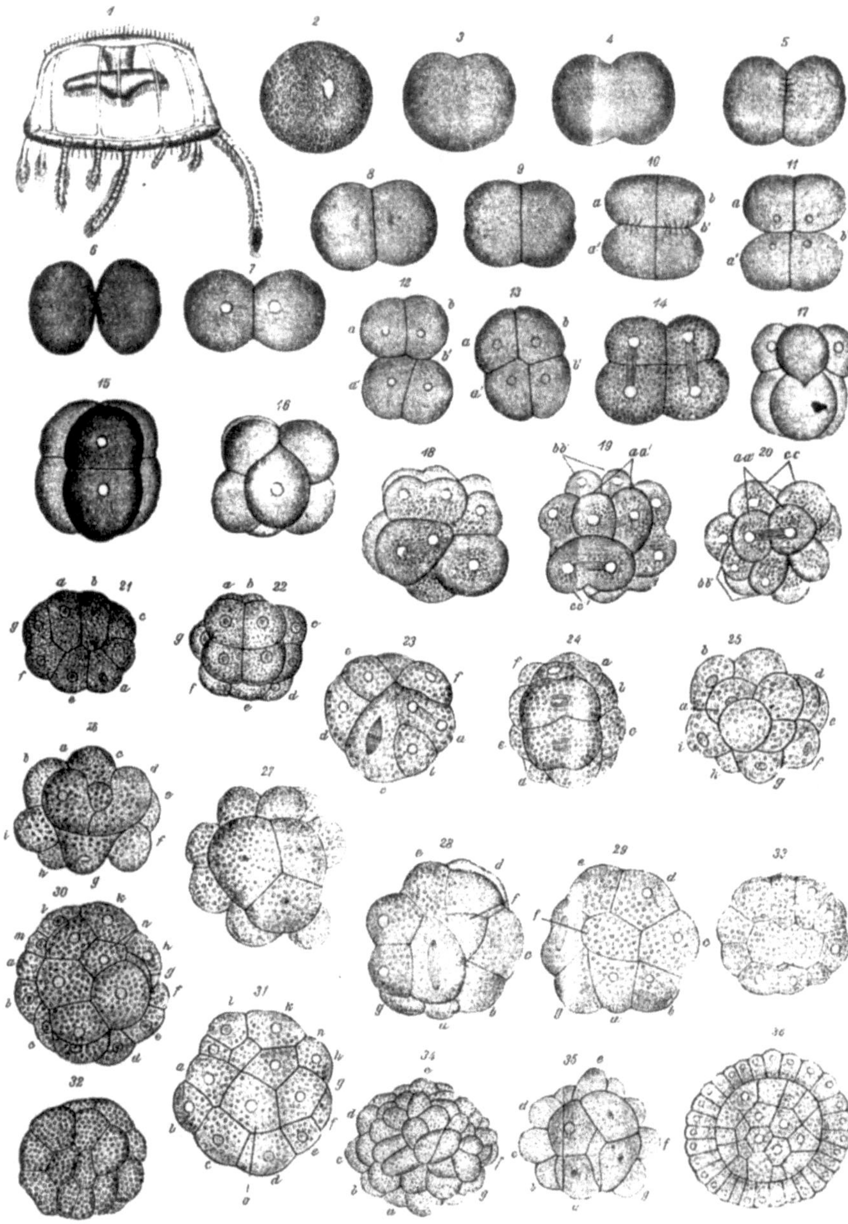

FIGURE VIII

Dr. 2–36 relate to *Polyxenia leucostyla.*

Dr. 6. Completion of first division. A live animal is depicted. 160:1.

Dr. 7. Two blastomers coming closer. A live animal is depicted. 160:1.

Dr. 8. Preparation to second division. A live animal is depicted. 160:1.

Dr. 16. 8-cell stage. A live animal is depicted.

nutrition for the ingested one; the latter may be regarded as a spore. In addition to small amoeboid cells with one ingested oval cell inside (Dr. 29), I found others with two or more ingested cells. Although I observed a number of such stages, I was never able to catch spores at the moment of division, so it is speculation on my part that they do indeed divide to produce two cells (Dr. 30). The latter either remain connected or incompletely divided. [.....]

The next stage (Dr. 31) shows four cells closely adjoined to each other and forming a regular cross-shaped figure; this symmetrical arrangement is useful in tracing the orientation of the cleavages. It shows that [the next] division occurs in a plane perpendicular to that of the first [two] divisions, and therefore this process replicates the early divisions in the eggs of medusae. It is more difficult to trace the following stages of division because they do not seem to form regular shapes. I frequently saw embryos with eight (Dr. 32), twelve (Dr. 33), and even more blastomeres in a group, in some cases in a radial arrangement. If the different embryonic stages are compared, both the embryo and its parts tend to show a considerable increase in volume, in spite of individual differences in size. [.....]

Further multiplication of embryonic cells leads to a stage resembling a blackberry; subsequently the cells become distributed in two layers (Dr. 34). This differentiation occurs gradually and is amazingly similar to the second delamination in the embryonic development of medusae and hydroids. In spite of intense search, I was unable to find any gastrula-like stage. [.....] Meanwhile, the embryo has grown bigger and cannot be remain hidden within the body wall any longer; it penetrates the gastrovascular cavity and remains attached to the wall at one point only (Dr. 36). When the ectoderm is covered with cilia, the embryos detach completely from the inner wall of the endodermal cavity [of the medusa]. They are close to the radial canals, and many of them leave the cavity and penetrate the canals, which are often completely filled with them.

The detached embryo is in the shape of a flattened corpuscle, with clearly visible embryonic layers of which the outer is significantly the thicker. A small cavity, initially completely closed, appears in the middle of the endoderm (Figure XII, Dr. 1). [.....]

As the lens-shaped embryo grows, its endodermal cavity increases in size significantly; its ectoderm changes into a thin layer of flat epithelial cells (Drs. 2, 12).

After completing the formation of the endodermal layer, the mouth breaks through in the center of the lower surface of the embryo; this aperture is surrounded by a buttress of significantly thickened endoderm. The following stages show considerable growth of the disk-shaped embryo. Firstly, two groups of stinging cells are joined by another two or, in some cases, even three. Small rudiments of tentacles, typically four, also grow in these areas (Dr. 4). Irrespective of the smallness of these organs, both embryonic layers participate in their formation (Dr. 14). [.....]

Thus a new two-layered embryo is being formed, by all accounts similar to an embryo developed from a spore. Since such buds can be found in medusae of all ages, it is reasonable to infer that the same embryo is capable of producing several independent buds that detach from the maternal body one by one, as in *Cunina rubiginosa*. [.....] We soon see that the larva of *Cunina* consists of two clearly separated parts: an upper disk with rudiments of tentacles, sometimes with a bud in the center, and a bubble-shaped body with a mouth aperture in the middle of its oral surface. [.....]

A closer look at the margin of the aboral disk shows four rudimentary tentacles, and sometimes five or even six (as in Dr. 5), with rudimentary tubercles on the lappets between them. [.....]

As the size gradually increases, the aboral disk overgrows the lower part of the larva, and the shape of the larva itself changes from globular to flattened (Dr. 6). [.....]

Flattening of the larval body, growth at the edge of the umbrella, and accumulation of mesogloea completes the metamorphosis of the larva into a young medusa. The medusa emerging from this metamorphosis is considerably smaller, and it leaves the maternal body and starts swimming as an independent organism. This animal, of diameter 5 mm and height 1 mm, drastically differs from *Cunina proboscidea*, even at first sight. Its body is not a shallow cone, as in the latter, but is shaped like a watch-glass; there is no sharp division between the disk and the edge of the umbrella (Drs. 7, 8). There are four rudimentary tentacles near the edge of the umbrella, at some distance from the periphery, but, since their development ceases at an early stage, they are almost invisible. The marginal bodies [statocysts]are more noticeable; their numbers vary significantly from animal to animal and also in different parts of the same animal. I have seen animals with 12–16 marginal bodies. [.....]

Free swimming medusae are completely developed sexually. The sexes are always separate; the ovaries, however, look very similar to tentacles. Each ovary is a ring-like thickening in the lower part of the stomach. [.....]

Medusae that developed as described survived for weeks in my laboratory without showing any further change; they did not develop eggs either.

Comparison of the generation of medusae originating through sporogony and budding with genuine *Cunina proboscidea*, they differ so drastically that Haeckel considered them to be two independent species. The original generation of *C. proboscidea* are true representatives of the Cunanthidae,[31] with gastric pouches, peroneal canals, radial canals with "otoporpae", and complex sexual organs, while the second generation, which originated from spores and buds, has a simple stomach sac, radial gonads, and no otoporpas, i.e. features of species of the Solmaridae.

This second generation is not the larva that will later metamorphose and mature into *C. proboscidea* , and this is shown firstly by the different composition of the marginal bodies; these are not as developed in the original generation, as they should be according to this point of view. On the contrary, they have a simple design and usually contain only three crystals [62], far less than the comparable number in the solmarid-like generation. In addition, young medusae of the first generation, which are no bigger than the free-swimming solmarid-like *C. proboscidea*, also turn out to be typical Cunanthidae [63], with otoporpas and a gastrovascular system [with pouches]. We are, therefore, left with only one conclusion, namely that the two generations are part of the developmental cycle of *C. proboscidea*,[32] and that one of them, with the exception of the marginal bodies, is far more complex than the other that originates from spores and buds.

Figure IX

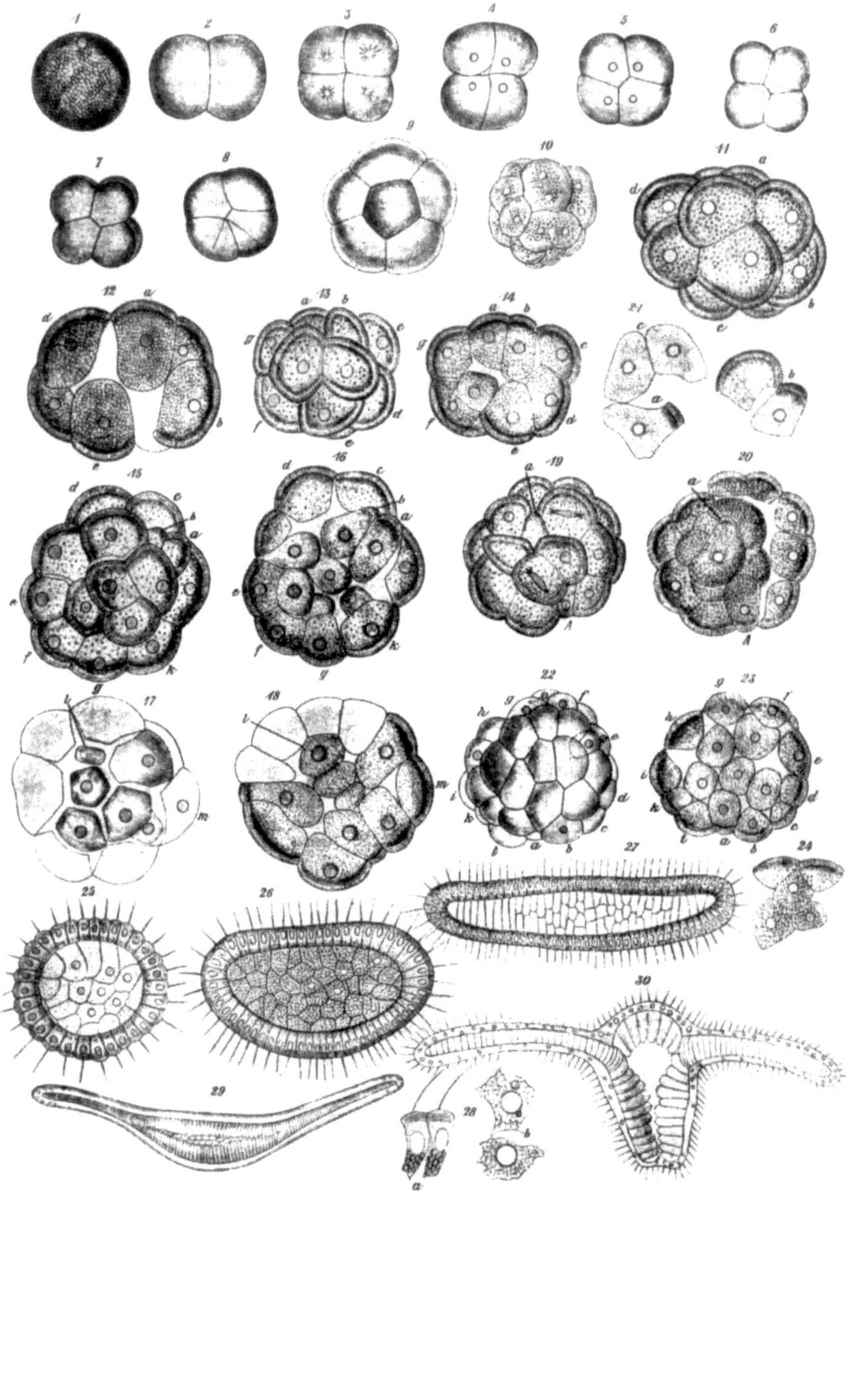

FIGURE IX

Dr. 1–30 relate to *Aeginopsis mediterranea J. Müll.*

Dr. 8. Egg in 25 minutes after third division. Osmium fumes, borax carmin, glycerin. 200:1.

Dr. 10. Beginning of fourth division. Osmium fumes, glycerin. 105:1.

Dr. 11. 16-cell embryo in optic section. Osmium fumes, glycerin. 105:1.

Dr. 12. The same embryo from the surface. 200:1.

Dr. 13. Another 16-cell embryo from the surface. Osmium fumes, boric carmin, glycerin. 180:1.

Dr. 14. The same embryo in optic section. 180:1.

Dr. 27. The following stage of larva. Osmium fumes, Bil carmin, glycerin. 100:1.

More research is needed to answer the question of what changes take place in medusae developed from eggs of both generations. [.....]

We can see a certain analogy between sporogony in *Cunina* and pedogenesis in larval *Miastor*, where again sexual organs and gametes are formed in the embryo. Only in the Coelenterata is sporogony completely isolated.

CHAPTER 7. PHYLOGENETIC CONSIDERATIONS

ON THE HISTORY OF THE THEORY OF EMBRYONIC LAYERS – HYPOTHESES ON THE ORIGIN OF THE METAZOA – PHYLOGENETIC SIGNIFICANCE OF THE FIRST THREE DIVISIONS – FORMATION OF ENDODERM – NOTES ON THE GASTRAEA, PLANULA AND PLAKULA THEORIES – DEVELOPMENT OF THE PHAGOCYTELLA THEORY – RESPONSE TO CERTAIN OBJECTIONS – RELATIONSHIP OF PHAGOCYTELLA TO TURBELLARIA ACOELA – NOTES ON THE DIVIDING BLASTOPORE THEORY – ON THE PHYLOGENY OF "POROSCHIZA" – PHYLOGENETIC [CONSIDERATIONS] ON THE FORMATION OF MESODERM

Some ideas that founded the initial principles of contemporary comparative embryology were already known at the beginning of our century. As stated by the German school of natural philosophy: "Evolutionem, quam prima aetate quod subit animal, evolutioni, quam in animalium serie observandam putant, respondere."

Von Baer [64] disputed this conclusion when he claimed that the embryonic stages of animals cannot be compared with adult forms, merely with their embryos. Ultimately, however, he had to admit that the distinction between these positions was not as great as it originally seemed.

A considerable advance was made by L. Agassiz when he spotted certain parallels between embryonic stages of contemporary animals and dominant stages in the geological history of the animal world, but he did not draw valid conclusions from this conformity.

After attempts to establish similarities between embryonic forms and adult animals and between embryos and fossil forms, Huxley, in 1849, equated the embryonic tissues of vertebrates and the major layers of coelomate invertebrates, and he called these layers ectoderm and endoderm.

This idea was not left unnoticed in England, and it was greatly popularized by Herbert Spencer in one of his wonderful essays, "On Social Organisms" [65].

Huxley's hypothesis had no followers for a long time; this resulted from reaction against any theory of embryonic layers, which [view] found its most vivid expression in Weismann's work on the embryology of dipterans. Weismann's ideas were clearly consistent with the "theory of types", prevalent at that time, according to which morphological comparisons are permissible within one type of animal only.

Attempts to explain anatomical similarities between the developmental stages of animals were [initially] inspired by the concept of some general design, which was understood only in an idealistic sense, but the reason for such similarities, as we have been taught by Darwin, is their genealogical kinship. Darwin paid particular attention to the embryo; the embryo is more consistent than the adult, thus allowing conjecture, with considerable conviction, on the form of its predecessors, and also the similarity between embryos is an indication of the common origin of adult animals. He attempted to confirm these general embryological resemblances with empirical data, and exceptions were treated by him as evolutionary changes resulting from heritable adaptations of the early stages. Darwin concluded that parallelism, established by Agassiz, was induced by the genealogical relationship of animals, and that parallelism took place only if no obscuring secondary evolutionary changes had occurred.

Fritz Müller, in his remarkable book *Für Darwin*, not only advanced Darwinian principles deductively but gave them greater analytical depth by using concrete examples from the embryology and phylogeny of lower organisms. He also showed how data from comparative embryology should be processed. F. Müller was quite persistent in stressing that phylogeny is repetitive, from a genealogical point of view, only when the progeny do not deviate from the maternal form, i.e. when they go all the way to this form and, after reaching it, evolve further. [.....]

After this fundamental work was published, and partly under its influence, intense experimental research on the embryology of animals was initiated, and lower animals became the object of major interest. In 1865, Koelleker, in the second part of his *Icones Histologicae*, [90] came, completely independently, to conclusions almost identical to those of Huxley. At the beginning of his work, he mentioned that correspondence in design between a hydroid and a young vertebrate embryo was so striking that further studies, combined with analysis of structure and histology of many animals, would necessarily lead to a simple law of evolution. This further research did not keep the spectators waiting. The theory of embryonic tissues, somehow forgotten with respect to invertebrates, was revitalized and enriched by such a substantial number of new facts that it soon became the basis for a new branch of embryology. The seminal effect of Kovalevsky's description of development in the lancelet [66] was certainly critical for this new development: in the embryology of this animal he revealed facts that brought together the development of vertebrates and invertebrates. After discovering the two-layered, ciliated lancelet larva, he turned to other animals in the search for analogous forms of development, and he accumulated many extremely valuable facts.

Researchers of this persuasion concentrated their efforts on the discovery and analysis of the major stages in embryonic development in order to compare them with the development of the lancelet, and were interested predominantly in lower animals with larvae of relatively simple design. I, meanwhile, turned to embryological phenomena in higher invertebrates to develop the theory of embryonic tissues. Firstly, I studied the embryology of *Sepiola* [67], where I found two embryonic layers and traced their participation in organ formation. In my search for embryonic tissues in the Arthropoda, insects did not give conclusive results, but I discovered such layers in the higher Crustacea (*Nebalia*) [68] and in *Scorpio* [69]. In this latter form, two embryonic layers were recognized first (1866), but soon after (in 1868) I found a third. My observations showed that the upper layer in *Scorpio* produces the neural cells of the central nervous system, while the middle layer divides into two, before the formation of segments and the merging of the cavities of several segments to form a general body cavity; the lower embryonic layer turned out to be the intestinal-

Figure X

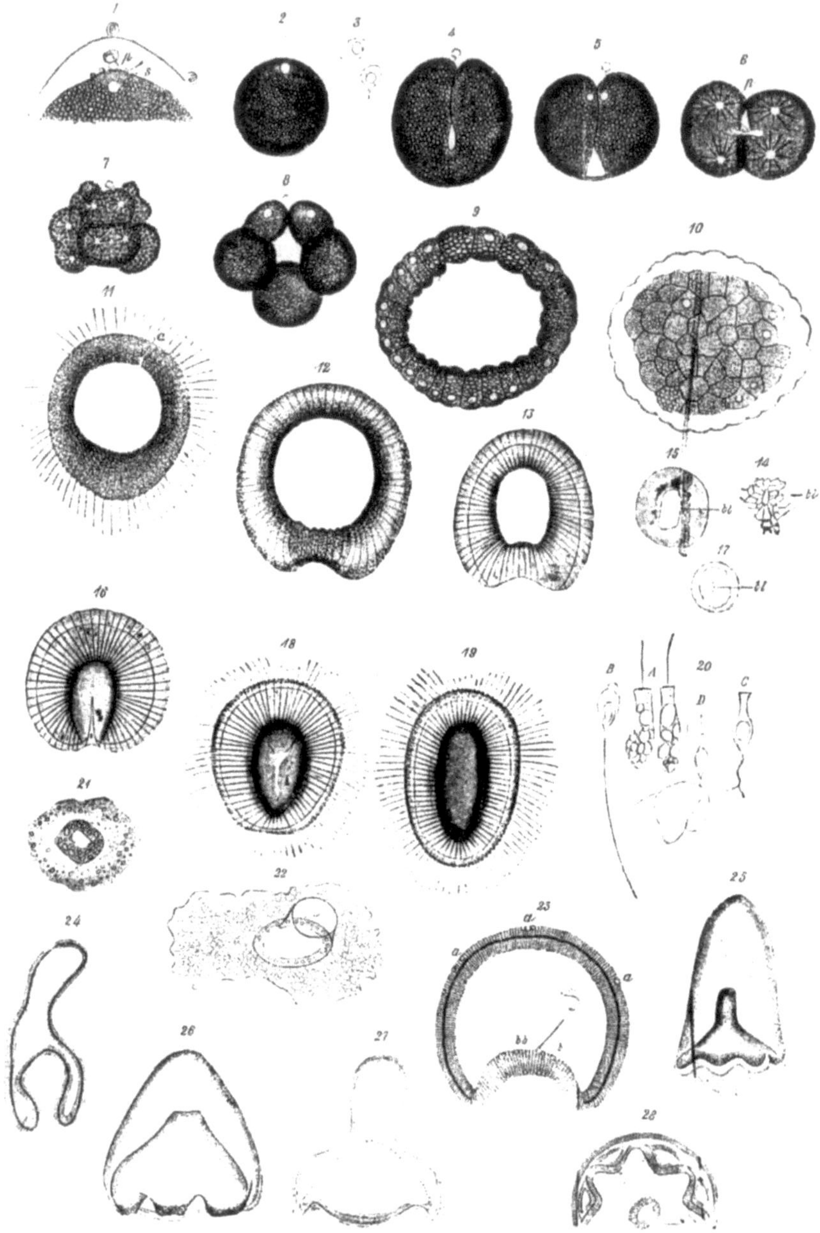

FIGURE X

Dr. 1–22 relate to *Nausithoe marginata Köll.,* Dr. 23–28 relate to *Pelagia noctiluca.*

Dr. 2. Newly laid egg. 105:1.

Dr. 5 Beginning of the second division depicted from upper pole. A live animal is depicted. 105:1.

Dr. 24. Longitudinal section of a more mature gastrula.

Dr. 25. 4-day old larva. A live animal is depicted.

Dr. 26. More developed larva.

Dr. 27. Longitudinal section of the same stage.

glandular one. These facts allowed me, in 1869 [70], to come to the conclusion that "the three layers in the scorpion embryo quite satisfactorily correspond to Remak's three layers in vertebrates". I insisted on this idea even though it seemed to me that the neural fibers emerged not from the upper layer but from the middle one. At that time, such a method of formation of the peripheral nervous system was also considered canonical for vertebrates.

Thus, in order to obtain sufficient data, studies in comparative embryology were undertaken from two different positions. Only after I had traced, in a generalized fashion, the embryonic layers of the scorpion, Kovalevsky [71], in his studies on the embryology of oligochaetes and insects, discovered the same three layers and traced them in detail. The revived theory of embryonic layers in invertebrates thus acquired a solid foundation, and a new branch of research on comparative embryology was founded, mainly by German and Russian scientists.

The very concept of embryonic layers was borrowed from the embryology of higher animals and transferred to invertebrates, and this "antigenealogical" method has resulted in a number of problems, some not yet overcome. Too often, while determining embryonic layers in doubtful cases, purely topographic features have been taken into consideration; for example, the so-called endoderm in the Orthonectidae and Dicyemidae[33] is merely reproductive cells within an outer body layer. Ed. van Beneden [72] considers topographic position as the only principle in defining embryonic layers. [.....] That is why Haeckel's attempt to formulate a hypothesis according to which the embryonic layers, or at least the first two, should be considered as primary tissues, is of the utmost importance. This move created a solid, and purely genealogical, foundation. From this point of view, a formation may be regarded as endoderm only if it combines several features of this primary tissue, and not if it merely coincides with it topographically. If, for example, the inner cellular mass of orthonectids were to perform digestive functions, it should, without doubt, be regarded as endoderm, but when, in reality, it consists exclusively of sexual cells, such an analogy cannot be accepted. The major difficulty relating to the genealogical interpretation is that it does not provide a solid foundation for solving the problem [of the derivation] of embryonic tissues, for we do not know the origin of multicellular animals. To come up with even the vaguest ideas on this topic, we should start with hypotheses consistent with known data. Just as evolutionary theory (*Descendenzlehre*) was transformed from pure speculation to a proven and reliable theory by demonstrating its ability to explain a significant number of relevant phenomena, hypotheses on the origin of primary tissues, to become a theory, should be tested on a considerable amount of empirical data.

Although in the contemporary animal kingdom any transformation from Protozoa to Metazoa is highly unlikely, some conjectures on this matter were considered. Two methods of transformation were suggested: [firstly,] through the differentiation of protoplasm around some cells in multinuclear protozoans, and [secondly,] by the integration of a number of individuals in a colony of protozoans to become a multicellular formation. In recent years, close affinity between [some] Ciliata and larvae of the Turbellaria has been postulated, leading logically to the concept of phylogenetic relationship and, further, to a hypothesis on the origin of the Metazoa.

This logical possibility was indeed proposed: many researchers raised their voices in favor of the idea that multinuclear *Opalina*-like infusorians had been transformed into multicelulars. Iering [74] and, more recently, Saville Kent [75] endorsed this position. From this point of view, the mouth aperture and poroshiza[34] of infusoria should be considered homologous to the same organs in multicellulars. [.....] If the whole of the developmental process is considered, the blastoderm of insects is most relevant to this hypothesis. S. Kent indeed refers to early embryonic stages in the Aphidae,[35] and this clearly shows that such a hypothesis is incorrect. The entire development of the lower Metazoa is virtually ignored in this presupposition, which can explain only the origin of the blastoderm in insects, i.e. in animals that undergo a secondary change. But even among these, there are forms that cannot be explained by this hypothesis. Consider, for example, the division of egg cells in *Podura*, an insect occupying the lowest position in its class, corresponding to myriapods. Because of these occurrences, homology between the mouth, poroshiza and vacuolar system of infusorians seems completely unsubstantiated. On the other hand, the hypothesis of the transformation of a colony of flagellates into a primitive metazoan can explain the majority of the most important facts. From this viewpoint, total division of the egg is primitive, and may be likened to the division of a flagellate at the start of colony formation. The flagellate-like character of the cells of many blastulas is explicable as inheritance from the Flagellata. Limitations of the described hypothesis, like its inability to correlate the mouth and other organs of the Protozoa with corresponding parts of the Metazoa, seems to be compensated by its ability to explain the origin of sexual reproduction, as first shown by Bütschli [76]. No wonder that the majority of embryologists, including R. Lankester and Balfour, supported this second hypothesis, which laid the foundation for further theoretical speculations.

It is quite natural to turn now to a detailed analysis of [the hypothesis that] a colony of Flagellata was the direct ancestor of the Metazoa. Bütschli [77] postulated that the Metazoa had two roots: colonies of Choanoflagellata, from which sponges originated, and colonies of Flagellata, which gave rise to other multicellulars.

Even if one accepts the factual foundation, which is too meager for [Bütschli's] risky claim, there are certain considerations that cannot be ignored, namely, that the two groups are unimaginably close to each other, and that the collar, considered as the main distinction between the two, can, in some cases, be completely absent. I shall discuss the relationship between sponges and flagellates later. Regardless of whether or not the Flagellata from which the Metazoa originated had a collar, it is undeniable that they could swallow particles of solid food, as indicated by the wide occurrence of intracellular digestion in the Metazoa. I cannot, therefore, agree with Bütschli, who considered the form of digestion unimportant in the origin of the Metazoa. [.....] Rather, I suggest that the further differentiation of the flagellate colony that led to the formation of the first multicellulars was never indifferent to the form of digestion. While some colonial flagellates developed plant pigments for the purpose of assimilation, others still kept their animal way of feeding, with some individuals specially adapted for the intake and digestion of food. That this conclusion is not merely speculative may be shown by a comparison of such colonial Flagellata as the Volvocineae and *Proterospongia*.[36]

Figure XI

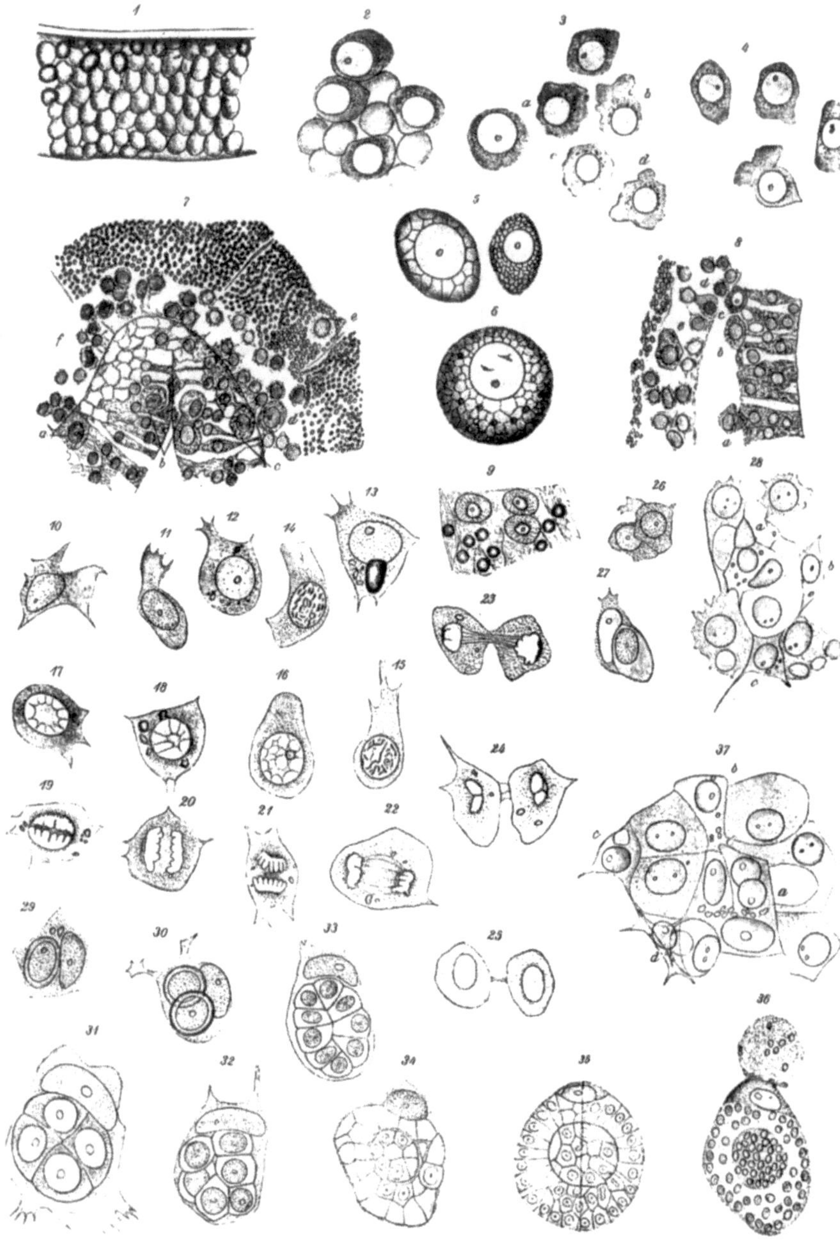

FIGURE XI

Dr. 1–37 relate to *Cunina proboscidea Metchn.*

Dr. 29. Amoeboid cell with a spore. Osmium. 870:1.

Dr. 30. Another amoeboid cell with a spore divided in two cells. Osmuim, Bil carin, glycerin. 870:1.

Dr. 31. Amoeboid cell with a 4-cell stage of spore division. Osmium. 870:1.

Dr. 32. Stage with 8 blastomers. Osmium, Bil carmin, glycerin. 870:1.

Dr. 33. The following stage with 12 cells of division. 870:1.

Dr. 34. Embryo with two leaves. Osmium, Bil carmin, glycerin. 540:1.

Dr. 35. The following stage. Treatment is the same. 350:1.

Dr. 36. Longitudinal section of the same stage. 540:1.

It became quite important for further theoretical review to acquire more accurate and more detailed information on the development of colonial flagellates. Sexual reproduction may be assumed because it is dominant in *Volvox*. As already pointed out, sexual reproduction is an important argument in favor of the origin of the Metazoa from the Flagellata. As to the facts of division, we know that contemporary flagellates behave differently in this respect. Genuine Flagellata predominantly divide longitudinally, although diametrical division occurs in some species, as, for example, *Phalansterium consociatum*, according to Tsenkovsky [78], and *Ph. digitalum*, according to Schtern [79].

Their close relatives, the Choanoflagellata, also have two methods of division. According to S. Kent, *Salpingoesa campanula* divides longitudinally [80], while some other species of the same genus [81], e.g. *S. inquillata* and *S. gracilis*, divide only diametrically. [.....] Because of the importance to the entire colony of how individuals divide, it is intriguing to speculate on the behavior, in this respect, of a hypothetical Metazoa-Flagellata.

As will be recalled from Chapter II, the first three divisions take place in all three directions and are identical in medusae with completely different developmental patterns: hypogenetic and metagenetic, regardless of different methods of endoderm formation. They are similar in completely different animals in which the blastomeres divide completely. Consequently the suggestion that ancestral metazoans had the same division pattern appears quite valid. Probability is added to this claim by the fact that embryos can originate in many different ways, as is borne out by both plants and animals with irregular patterns. Even the formation of a blastula can occur without the first three divisions. We see that in *Volvox* all divisions are meridional, which leads to a lamellar embryo resembling *Gonium*; there is no cleavage cavity, and cleavage furrows are formed only later, after the movement of platelets of cells to one pole [82].[37]

If we now consider the cleavage process in evolutionary terms, the reasoning cited above seems irrefutable. An analogous process in [another branch of] the organic world is clearly shown by the process of division in *Dendrobacterium*. While the majority of forms divide diametrically, there are rare exceptions of species with longitudinal division, such as the branching species *Dendrobacterium ocellatum*, a parasite of *Daphnia pulex*, which I discovered.

This bacterium with only one mode of division is supplemented by others in which cells divide in two meridional planes (as, for example, the micrococcus of gonorrhoea), and still others (like sarcinas) with three-dimensional division, which brings them close to the majority of the Metazoa, as well as to the hypothetical Metazoa-Flagellata, in which there was complete division. Since a cleavage cavity typically appears just after the third division, and the embryo transforms into a blastula quite early, it is quite possible that the ancestors of the Metazoa also came together as something similar to flagellate colonies.

If the features of the metazoan-flagellate described above are accepted as a basis for further discussion, some light, I suggest, can be shed on the origin of the primary tissues. We saw in Chapter 3 that endoderm formation in medusae occurs in quite different ways. Recalling these methods of formation, it was established that endoderm can either appear in many areas of the embryo simultaneously, i.e. *multipolar*, or in just one zone, i.e. *hypotropic*. The multipolar mode of endoderm formation may result from: (1) multipolar migration of cells from the surface to the interior of the embryo; (2) primary delamination, through the diametric division of blastoderm cells; (3) secondary delamination, which takes place after the completion of the entire process of morula formation; (4) mixed delamination, in which endoderm cells are formed partly by diametric division and partly by migration. Some of these methods of formation cannot be strictly separated from each other. Hypotropic or restricted endoderm formation can take place either as migration of blastoderm cells in the lower part of the larva, or as genuine invagination. This mode also embraces the peculiar form [termed] epiboly, a variant of irregular division.

The questions that arise are: how can these different modes of formation be explained, and what theory of the origin of the primary tissues is most compatible with this explanation?

It is quite obvious that any theory of the origin of the Metazoa from multicellular Protozoa (infusarians and radiolarians) that leaves unexplained such features of embryonic development as migration from the surface, primary delamination, and invagination, should be rejected without any further evaluation.

The gastraea theory, as is well known, is very instrumental in solving problems through reduction of many developmental processes to primary invagination; it is quite helpful to the understanding of complicated procedures of endoderm formation, especially when vertebrates are concerned. At the same time, it gets into serious difficulties over delamination, and, from the first, this was admitted by Haeckel in his first attempt to formulate this theory. "The biggest challenge to this explanation," wrote Haeckel in his monograph on calcareous sponges (Vol. I, p. 467, note), "is presented by the fact that the gastrula emerges from a morula in two completely different ways: by creating a central cavity in the morula and subsequent splitting outside the digestive cavity formed in this way, and by the formation of an embryonic capsule, which is a hollow sphere with walls of a single layer of cells, and the invagination of this capsule inside itself."

Haeckel considered it possible to overcome these difficulties by suggesting "secondary modification of ontogenesis". In his major work [83], he frequently reiterated the assertion that delamination, if it indeed takes place in the animal kingdom, is a cenogenetic process "which only secondarily evolved from the paleogenetic process of invagination". He, however, did not elaborate on his explanation of this falsification. This is quite puzzling because Haeckel himself considered such an explanation to be a major problem for his theory. For a long time, Haeckel and his school (namely the Hertwig brothers) doubted the very existence of delamination, but now, after O. Gaman [84], one of Haeckel's followers, observed delamination in hydroids (after descriptions by many researchers, including B. Ollman, F. E. Schultze and myself), they finally accepted it. He attempted to overcome this difficulty by viewing the non-laminated planula as a true gastrula and postulating its evolutionary descent from an invaginated gastrula. "I think we can talk about a gastrula even, as seems quite possible, in the absence of both formations; the blastopore and digestive cavity are of secondary importance. That is why I consider a

Figure XII

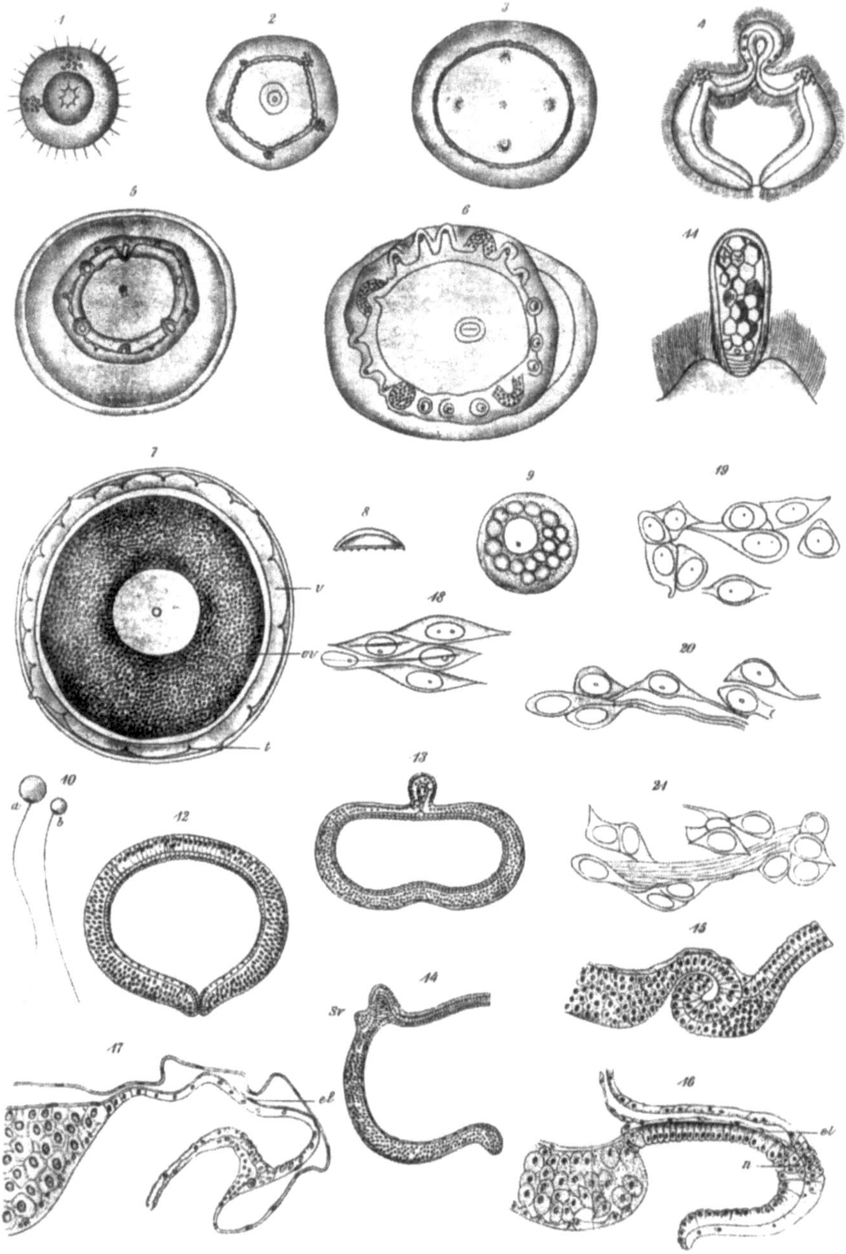

FIGURE XII

Dr. 1–21 relate to *Cunina proboscidea.*

Dr. 1. Sporic larva without a mouth. A live animal is depicted.

Dr. 2. More mature stage of larva with a mouth and a spore (cilia are not depicted). A live animal is depicted. 105:1.

Dr. 5. Larva with a pillow-like sexual embryo. A live object is depicted. 41:1.

Dr. 6. Similar stage. A live animal is depicted.

Dr. 7. Sexually matured female medusa. *O* ovary, *v* – velum. A live animal is depicted. 13:1.

Dr. 8. The same medusa in profile. 3:1.

Dr. 10. *A* – sperm of major generation. A live animal is depicted. *B* – sperm of second generation. A live animal is depicted. 540:1.

Dr. 12. Longitudinal section of larva with a mouth opening. 160:1.

Dr. 14. Half of longitudinal section of a more mature larva. *Sv* – embryo of subumbrella and velum. 105:1.

planula as resulting from gastrulation by delamination." [85]. These claims do not facilitate understanding of how invagination (emboly) can change into delamination or the part played by cytogenesis in its origin. If, in the meantime, we take into consideration that invagination is concentrated at one end of the embryo and is limited, in medusae at least, to a relatively small area of blastoderm, while primary delamination or multipolar ingression takes place in different parts of the embryo, it becomes clear that transformation of the first method into the second would encounter insurmountable difficulties. It is easy to imagine how delamination, initially restricted, could gradually spread over half the blastoderm, as is frequently observed in different animals, or how an embryo consisting of a single layer of endoderm could change into a ball of cells linked to each other that later becomes covered with ectoderm, but it is quite a challenge, seemingly unrealistic, to derive a scattered assemblage of endoderm cells, which do not touch each other but alternate with ectoderm cells (as in the development of *Aeginopsis*), or originate as inner parts of blastoderm cells, from a simple invagination.

It is just possible to attempt to derive multipolar ingression from multiple invaginations, with each archenteron consisting of one cell only! But the very formulation of this suggestion is enough to show its complete inadequacy. This type of explanation also leads to an impasse because it cannot explain primary delamination. In a further attempt to link the explanation of gastrulation in the Craspedota with epibolic processes occurring as variations in the development of *Polyxenia leucostyla*, one should be reminded that such a formation is closely associated with uneven distribution of blastomeres. Even when cell division occurs in *Polyxenia* it produces mixed delamination, a form that cannot be derived from invagination, instead of genuine invagination, which in other Metazoa results from equal cleavage in an amphiblastula. This example shows that epiboly in *Polyxenia* should be considered rather as a variant of mixed delamination, which, in this case, evolved secondarily as a result of unequal cleavage. Under these circumstances, any further discussion on affinity between invagination and secondary delamination becomes redundant, for it cannot lead to any meaningful results. It is far more natural to accept the existence of a link between invagination and hypotrophic ingression, as already stated by Klaus and others, but, nevertheless, a preliminary discussion on related questions appears to be crucial to determine which form is primary.

The impossibility of explaining multipolar formation of endoderm should be regarded as the strongest objection to the gastraea theory, but there is also a list of other problems. [The theory] was proposed when the idea of cavital digestion was dominant and universal and when intracellar digestion in lower animals was not yet known, but it is not in accord with contemporary physiological ideas. According to the gastraea theory, there is a genuine abyss between the one-layered blastaea and the two-layered gastraea with a digestive cavity; this abyss, however, can be easily filled if we attempt to explain the formation of endoderm from another viewpoint. Further complications that follow from the acceptance of a homology between all known forms of gastrula seem not worth discussion since they do not relate directly to the question of the origin of endoderm. They will, however, be discussed later.

The planula theory of R. Lankester [86] results directly from events observed in the development of *Geryonia*, and considers transverse division of blastoderm cells as the primitive method of endoderm formation. Even if we accept Lankester's suggestion, which replaces invagination with primary delamination, it is still impossible to explain such cases as multipolar ingression where there is no delamination at all, yet endoderm still appears (formed by longitudinal division of blastoderm cells), and it moves to the interior of the embryo from different parts of the surface. The importance of this type of formation is reinforced by hypotropic migration, which too is based on the longitudinal division of blastoderm cells. There is, moreover, another objection which should be leveled at the planula theory, and the gastraea theory also, namely that these two theories are based on the acceptance of cavitational digestion in lower animals, and thus contradict contemporary physiological data. R. Lankester claims that the formation of a digestive cavity into which gastric juice is discharged occurs prior to the development of endoderm, and that, therefore, the inner parts of blastoderm cells served as digestive elements when the wall was still one-layered.

All these suggestions turned out to be completely unacceptable once it became known that intracellular digestion takes place in many lower organisms and can even be found in some mollusks (e.g. *Phylliroe*).

While the gastraea and blastaea theories are based on a blastula consisting of similar cells, Balfour [87] takes the amphiblastula[38] as the basis for his views on the change from primitive to multicellular organisms; consequently, his opinion may be termed the amphiblastula theory. It should be regarded as a modification of the gastraea theory, and therefore all the objections to the latter are relevant. [.....] Regarding the applicability of Balfour's theory to sponges, its inadequacy springs from the fact (observed earlier and more recently confirmed by many researchers including me) that show that, in some sponges (namely calcareous sponges and Halisarchinae), the ectoderm is the body layer that catches food. These facts were reconfirmed recently for *Oscarella lobularis* by Geider [88]. He also refutes Balfour's arguments that the sponges occupy an isolated position among the Metazoa.

The amphiblastula theory is closely related to the plakula theory, developed by Bütschli [89]. This is not merely because Bütschli supports the idea of a separate origin for the sponges, but, much more importantly, because a plakula may be considered, in many respects, as a compressed amphiblastula. Bütschli, recognizing the shortcomings of other theories on the phylogenetic origin of the primary layers, and [also of] attempts to combine invagination and delamination in a purely schematic and general way, derives both modes of endoderm formation from changes to the primitive plakula. Refusing to accept a ball-shaped colony of protistans, which was the starting point for previous theoretical reconstructions, Bütschli considers the primary form to be a one-layered *Gonium*-like plate, called in my later exposition a proplakula. [.....] While a plakula, which acquires a tube-like shape, changes into a gastrula, a proplakula, by means of a secondary delay in cell division, turns into a delaminated blastula. The assumption for such a suggestion is the major differences between shapes of blastulas, which may be considered either as inflated plakulas or as ballooned proplakulas. In the life-history of contemporary animals, a plakula is represented by the flattened blastula of *Cucullanus* (according to Bütschli),

Rhabdonema (according to Goette), *Lumbricus, Chiton* (according to Kovalevsky), and *Ascidia mentula*, and among adult animals by [the mesozoan] *Trichoplax adhaerens* F.E.Sch. Bütschli, however, does not pay sufficient attention to the fact that the flattened blastula of these metazoans corresponds to his plakula only in superficial shape, and that the method of reaching the two-layered state is not that followed by most plakulas, i.e. not by division "parallel to the plane of the plate". The plakula-like stage in *Phoronis, Ascidia* and other animals mentioned above results from flattening of the primary, more or less bubble-shaped blastula, but not vice versa, as is demanded by the theory. [Bütschli] likens delamination in the Geryonidae, resulting from transverse division of blastomeres, to plakulas and amphiplakulas of other animals; these latter should show transverse division of blastomeres, but this never happens in reality. Endoderm cells of delaminated blastulas do not arise from ectoderm cells above them but from cells situated next to them. We can conclude, then, that a plakula stage played no part in the whole history of development of contemporary animals with regular division. There is even less ground for regarding as a plakula that stage in the embryonic development of Ctenophora in which an interrupted layer of ectoderm occurs above the endoderm in a lamellar embryo [90]; this should scarcely be regarded by anyone as a primary stage on which to base morphological generalizations.

Turning back to developmental processes, there are objections to regarding the meridional division of the first four blastomeres leading to an eight-celled stage as a transient plakula stage. It does not explain anything, for two reasons: firstly, the eight-celled stage in the Geryonidae, which delaminate, is directly comparable to one in the Acraspeda, which invaginate, and, secondly, delamination through meridional division cannot be related to the plakula stage because this [flattened] stage has gone already.

Even if we admit that the plakula theory adequately explains the formation of an invaginated gastrula (as the result of meridional division in the proplakula), such occurrences as the formation of endoderm by multipolar or local ingression following longitudinal division of embryonic cells are outside its coverage.

According to Bütschli's theory, animals with delamination should have the same mode of blastula formation as *Volvox*. i.e. creating a blastula from the previous proplakula stage (resembling *Gonium*). Bütschli thinks it possible to accept the occurrence [of a proplakula] in *Geryonia proboscidalis* (accepting to Folle's data: see p. 243 of his book, cited above). In fact, the sixteen-celled embryo of Geryonidae is already a typical blastula, arising from the eight-celled stage in the same way as happens in medusae with different modes of endoderm formation. The eight-celled embryo, in its turn, is formed by equatorial division in all medusae and in most metazoans with equal cleavage. Among the latter, the greatest similarity to *Volvox* (in stages preceding the blastula) is seen, according to F.E. Schultze [91], in *Sycandra raphanus*, solely because this sponge has a lamellar eight-celled stage. This, however, does not support the plakula theory: it is simply that an invaginated tube is a feature of *Sycandra*.

The morphology of *Trichoplax adhaerens* F.E.Schultze[39] [92] is intriguing in many respects, but its relation to a plakula cannot yet be discussed intelligibly. The significance of the different tissue layers of this animal cannot be assessed even approximately; there are insufficient histological differences between the two body layers to form a conclusion on their relations to the embryonic layers [of other metazoans]. (A comparable situation occurs in sponges where the same embryonic layer, the endoderm, can be either flagellar epithelium in the [flagellated] chambers or lamellar epithelium in the collection chambers, as, for example, in *Oscarella lobularis*, according to Geider.) Such difficulties escalate, not only because of limited knowledge of developmental processes but also due to the absence of any data for determining the physiological significance of the body layers in *Trichopolax*. Thanks to the courtesy of Prof. F.E. Schultze and Prof. Klaus, I studied *Trichopolax* in 1883 in Gratz and Vienna, and was able to confirm the histological observations of the former [Schultze] in their entirety. My studies on the nutrition of *Trichopolax* gave negative results because these animals swallow no food at all, so the impression that they are adapted to a liquid diet was quite convincing.

Bütschli considers the plakula theory to be more sound physiologically than other views considered by him. [.....]

Referring to all that has been said above on the theories under discussion, we see that that they cannot unify the entire complex of embryological material, and also that they are crippled because of the lack of physiological explanations. Consequently another approach should be made. My studies on sponges [92] led me to make some very tentative remarks which seem to fit the known data on endoderm formation in the lower Metazoa and recently acquired knowledge on intracellular digestion. It was my understanding that the endoderm did not make its first appearance as a tubular enteron with a terminal aperture, as seen in a gastrula; it underwent a long process of gradual development leading to the formation of a solid parenchymella with intracellular digestion. This latter developed gradually, with surface blastoderm cells migrating to the interior of the blastocoel.

Finally, I had the idea of a two-layered parenchymella, which then transformed into a gastrula through shortening of the embryonic process and progressive differentiation of the digestive apparatus. Because at that time (1879) it was impossible to refer to any of the more or less highly organized flagellates with animal nutrition, I looked for support for my point of view in the reproduction of *Volvox*. [.....] Soon after that, in June 1880, S.Kent [94] discovered an exceptionally interesting form of flagellate which he introduced to science under the name of *Proterospongia Haeckeli.* Some animals in this colony of individuals that were distributed evenly and peripherally at first acquired an amoeba-like shape (see Dr. 1a) and penetrated the jelly-like mass that keeps the members of the colony together. The amoeba-like individuals remained there and started division and other changes which Kent regarded as sporulation (see Dr. 2). It is difficult to decide now, due to lack of sufficient knowledge, whether the change to an amoeboid form and penetration into the jelly-like mass was related to digestion. Because sporulation is still largely unknown, I would speculate that the granules so frequently observed by Kent could be either reserve digestive materials or swallowed food particles. It would be extremely interesting to examine species of *Proterospongia* (and also another species described by Oxley consisting of many individuals) with particular emphasis on digestion and reproduction. In any event, it must be granted that this collared flagellate has two types of individuals, flagellate and amoeboid, which can transform naturally from one to the other. The amoeboid form occurs in various surface areas and can penetrate the jelly-like mass. *Proterospongia* shows significant similarity to some two-layered sponge larvae, e.g. to the larva of *Aplysina sulphurea*, described by F.E. Schultze [94]. It is also quite obvious that differentiation of the amoeba-like individuals can be viewed as a first step towards endoderm formation.

In any case, I think that the peculiarities of *Proterospongia* are much easier to reconcile with my view, which some authors call the parenchymella theory, than with any [other] theory described above. Yet how can we correlate the parenchymella theory with data from developmental histories in general and the above-analyzed embryology of medusae in particular? To shed light on this, I refer first to what I regard as a logical consequence of events in the reproduction of the hypothetical metazoan-flagellate.

Having considered the universality of the first three cleavage divisions, which occur in the three spatial planes (sagittal, frontal and equatorial, and are, therefore, meridional and latitudinal), I concluded that the ancestors of the Metazoa divided in the same planes. Gradually these planes became fixed in such a way that some forms reproduced, predominantly or solely, through latitudinal division, while others, related to them, adopted meridional division. That similar events indeed did take place is shown by the species of Solpingoeca[40] described above. We have to accept, therefore, that some surface cells in our hypothetical metazoan-flagellate changed to an amoeboid form and permeated the colony in exactly the same way that we now see in *Proterospongia*; that, simultaneously, others divided latitudinally, the inner product of the division moving into the colony; while still other peripheral cells kept their surface position. Dr. 3 illustrates this schematically. (It is quite possible that, together with latitudinal and longitudinal divisions, other forms took place, and that some divided diagonally, as still occurs in some primitive organisms.)

This dual method of endoderm formation (through the ingression of rounded cells interlaced with the inner products of division, e.g. in *Polyxenia leucostyla*) is, in embryological terms, mixed delamination (see Dr. 4). While meridional division is usual in some forms (see Dr. 5), others generally divide latitudinally and are thus distinguished by the migration of rounded surface cells (see Dr. 6). Mixed delamination, therefore, gave rise to primary delamination on the one hand, and to multipolar ingression on the other. Secondary delamination should be regarded as merely a slight modification of mixed [delamination], the only difference being the slightly delayed indication of the distinction between the two types of cells.

It should certainly be accepted that multipolar endoderm formation is more primitive than the concentric method [i.e. invagination] (see Dr. 7); otherwise considerable difficulties occur, as we saw above. The change from multipolar to hypotropic ingression is, on the contrary, much easier to imagine, especially because the latter occurs predominantly in very active blastulas, while the former takes place in immobile or slow-moving embryos. Having considered endoderm formation in metagenetic medusae, I emphasize that latitudinal division is the rule, meridional division a very rare exception.

If restricted ingression can be derived easily from multipolar, it seems quite easy to derive invagination from restricted ingression. We saw in Chapter 3 that endoderm formation in *Laodice cruciata* differs from the same process in other metagenetic Craspedota by transparent cells formed at the rear of the blastula (see Figure IV, Drs. 24–28) which gradually move into and permeate the blastocoel (see Dr. 8). This leads to the formation of parenchymatous endoderm; due to the formation of a cavity, this endoderm acquires the appearance of an epithelium. The initial stage, with transparent cells, shows extraordinary similarity to the blastula of *Nausithoe marginata*, which is also characterized by differentiation of a hypotrophic layer of cells (Figure X, Dr. 12). If we consider further shortening of the development cycle, we can easily imagine endoderm cells which differentiate on the surface, then, instead of migrating individually, invaginate together and form an endoderm cavity directly (see Dr. 9). After the invagination of cells that have already differentiated, this process spreads to neighboring parts of the blastoderm, as a result of which the invaginating sac gradually increases in size. Further shortening of development may be envisaged, but it apparently did not occur in medusae. One may regard it as early differentiation of endoderm cells, so that even in the blastula all cells destined to invaginate may be recognized as distinct elements. Lamellar blastulas with two or more similar embryonic layers (e.g. as in *Lumbricus, Ascidia*, etc.), together with other [forms of blastula], should be considered as resulting from such abbreviation. It is quite obvious that further steps in the early differentiation of endoderm would lead to an amphiblastula; this may be recognizable in the early cleavage stages, and also by the structure of the egg.

If an amphigastrula evolved from an archigastrula as a result of early differentiation (which appears to be common knowledge since Kovalevsky's study of *Euaxes*), it appears certain that a similar form might have resulted from irregular cleavage and mixed delamination (as we observed in *Polyxenia leucostyla*). The example of *Polyxenia leucostyla* teaches us that an amphigastrula can develop in two different ways from two different early stages, and shows the polyphyly of this embryonic form.

From this point of view, endoderm should be regarded as the sum of cells initially formed from a blastula through migration or segregation and subsequently forming a conglomerate of unevenly developed amoeba-like elements. The archenteron and blastopore should be regarded as later acquisitions, conserved by shortening of embryonic development, and resulting in a definitive method of gastrulation.

Balfour [95], who spoke against the parenchymella theory, acknowledged, nevertheless, that it "conforms to the ontogenesis of the lower Hydrozoa quite well". Now, with the availability of new data, this conformity is even more apparent. However, when Balfour calls the transition from protozoan to multicellular state implied by the parenchymella theory as "impossible", I cannot argue because he does not substantiate his opinion. Much more cogent objections related to physiological aspects are presented by Bütschli. [.....]

When I published my migrational theory of endoderm formation [96], I initially referred to *Volvox*, in which proliferating individuals move from the surface of the colony to the interior. Now I can cite additional evidence from *Proterospongia*, in which the inward migration of individuals is clearly beneficial to the entire colony. In addition, some *a priori* considerations may be helpful. Further differentiation depends on some disparity between members of the colony: while some individuals generally swallow food, others are more adapted to moving and attracting food. In motile colonies, it would certainly be important to have individuals supplied with food, and thus heavier, not at the periphery but as near as possible to the center of the colony. Another benefit arises when the feeding individuals are placed in an environment suited to this function [feeding]. [.....] It is easy to imagine that feeding individuals would reproduce more frequently, thus establishing a correlation between intensive feeding and reproduction. Yet another reason for [inward] migration is the limitation on the spatial expansion of colonies, and this is particularly important for swimming colonies. Since any increase in the number of individuals in a colony enhanced its activity, it was important for cells that could not find room for themselves at the surface to adapt to existence inside the colony. For a long time members of the same colony probably differed only quantitatively: locomotor cells caught food particles with their flagella and swallowed the smallest of them. Even today, some ectodermal cells in the Coelenterata can be observed incidentally swallowing food [97]. Inner amoeboid individuals, however, are capable of swallowing larger particles missed by the locomotor cells. Perhaps active amoeboid individuals almost reached the periphery and caught particles that came into contact with the surface through multiple surface pores.

As analogous observations, one may cite the capture of external particles by amoeboid cells in hibernating *Halisacra pontica* which lose their flagellated canals, as well as ingestion by mesodermal cells in many lower animals. Gradually this type of differentiation became more advanced, locomotor cells losing their capacity to capture food, which [activity] was concentrated in the amoeboid phagocytes. Incidental fine pores between locomotor cells were able to grow larger and evolve into numerous foramens in the surface of the sponge. With the general specialization of individuals making up the colony, the next stage was the differentiation of surface individuals into *ectoderm* or *kinoblast,* while the amoeboid inner individuals became *phagocytoblast, parenchyma* or *mesoderm*. Cells of the latter group, when they were unable to capture larger food particles, would surround themselves with *plasmodium* – something quite similar takes place today in the endoderm of siphonophores and the mesoderm of many animals. Multicellular animals with the two primary tissues enjoyed increased activity and their feeding requirements were satisfied; it is quite possible that larger plant and animal organisms were also devoured. To augment this function, two or more inlet apertures could appear, subsequently leading to the formation of a mouth.

I can cite several analogies to the above-described differentiation of homogeneous animals to locomotor organs and phagocytes: in lower Coelenterata, as for example in hydroid polyps and some medusae (*Oceania, Phialidium, Cunina*), the entire enteroderm wall of the gastrovascular system is evenly adapted to consuming nutritive particles by intracellular digestion; all the endoderm cells are phagocytes, in addition to locomotor elements when they are furnished with flagella. In other Coelenterata, as for example *Aglaura* among Craspedota, Siphonophora, and

Ctenophora, the endoderm, by contrast, differentiates into amoeboid epithelial cells, the only ones that consume food, and flagellar epithelial cells, which serve to produce a water current and are unable to ingest any foreign bodies. A totally similar situation can be seen in sponges: usually flagellate endoderm cells behave like phagocytes, but in some siliceous sponges such a function is transferred wholly to amoeboid cells while flagellate endoderm cells serve solely to produce a water current.

It is self-evident that the migration of individuals from the surface of a colony into the interior (filled, it seems, with a jelly-like substance) does not preclude the meridional division of other individuals. (Also, when differentiation has already taken place, a peripheral product of division may be left in place, while an offspring may reappear, as a phagocyte, in a central position, within the colony.) The transitional form between the Flagellata and multicellulars, from which the latter descended, I initially called parenchymella. Now, I should like to rename it *phagocytella*, because the word reflects an important attribute of this form. My hypothesis is that phagocytella possesses two primary tissues, kinoblast and phagocytoblast, which, however, were not as distinctly separated from each other as the embryonic layers of the majority of the Metazoa; it seems that the replenishment of phagocytoblast from inwardly migrating kinoblast cells took place for quite a while. Regarding the sequence of development, it may be stated that the eggs of phagocytella (which reproduced sexually) underwent regular cleavage, with blastomeres dividing in three-dimensional planes, and that a blastocoel developed early and then was filled by migrating cells together with the products of cell division.

The phagocytella theory is consistent with our knowledge of the morphology and physiology of sponges. The embryology of different sponges shows a variability almost exceeding that in medusae. Although they have not been sufficiently studied, it is clear that, in the embryological development of these animals, gastrulation occurs both by [cell] migration and delamination.

All that has been said above on the embryology of medusae is, in general, applicable to the various ways of endoderm formation in sponges. The primary tissues of the latter remain in a relatively primitive state, so extending the concept of embryonic tissues to the layers in sponges encounters significant difficulties, as I have already pointed out. Because "endoderm" often transforms into "mesoderm" (suggesting that they are closely linked [98]) I think that both layers together should be regarded as phagocytoblast. The fact that cells that form the phogocytoblast do not appear simultaneously but through gradual migration from the endoderm (e.g. in *Halisacra*) does not undermine this view. There is nothing in the basic structure of sponges which proves their dissimilarity from other Metazoa, as Balfour and Bütschli attempt to argue. On this matter, I completely support the majority of spongologists, including K. Geider [99].

The absence of a mouth aperture together with the presence of many incurrent pores can be considered as significant differences only when these formations are misinterpreted phylogenetically. On the contrary, these peculiarities of sponges are easily explicable in terms of the phagocytella theory. The dominant role of amoeboid cells as ingestors of food would seem strange only if we could talk of differentiated mesoderm with respect to sponges; but because the sponges also possess phagocytoblasts, which should be considered as embryonic endo-mesoderm, the above-mentioned fact turns out to be straightforward and clear.

Just one year after I published my ideas in "Spongiologische Studien", Balfour came to conclusions on the development of the gastral cavity[41] which agree with the major points of the phagocytella theory [100]. [.....] In view of this, it seems quite surprising that he decisively supports the gastraea theory and considers the parenchymella theory quite unacceptable.

Recently, Goetthe [101] restated the view that there is close affinity between the gastric cavities [of sponges] and Turbellaria (Acoela), and also that the endoderm in primitive metazoans is parenchymatous, without, however, any reference to Balfour or myself. One could conclude from this that a number of scientists have been forced to accept the occurrence of a phagocytella-like stage in development. Although the latest studies on the anatomy of Acoela (by Kleinenberg, Pereyaslavtseva, I. Delazh, and me) correct Graff's data on the absence of mesoderm and a neural system in these animals, they do not challenge the fact that the digestive organs of the Acoela are of a very primitive type. Recently it was claimed that the above-mentioned worms possess a definite digestive cavity; I have to maintain that genuine Acoela possess endodermal plasmodium with a vacuole of variable size that can simulate a digestive cavity. These facts were investigated by me in a transparent pelagic mass [of acoelans] in Messina (where they were studied on numerous occasions by Kleinenberg), and I was convinced of the complete accuracy of my conclusions. The embryological studies of Pereyaslavtseva [102] and Repyahov [103] suggest that, in the Acoela studied by them, a gastrula stage follows cell division, and the second of these authors [104] claims that this is a basis for regarding the Acoela as degenerate worms. My objection is that the formation of a gastrula should not necessarily be considered as a primary phylogenetic process. We saw, in the embryology of medusae, the polyphyletic evolution of gastrulas from various primary formations: an archigastrula in the case of *Nausithoe* and *Pelagia*, an epibolic gastrula in *Polyxenia leucostyla*; its formation, in the process of phylogenesis, can easily be conceived as a secondary embryonic adaptation. In addition, it should be noted that any final conclusion on [the occurrence of] a gastrula in the Acoela can only be reached when the above-mentioned studies of Pereyaslavtseva and Repiahov have been published in their entirety.

Because, in my view, a gastrula can appear independently in embryonic development, many difficulties of the gastraea theory can be eliminated. I refer the reader to my "Comparative Embryological Studies" [105], in which I attempt to shed some light on this aspect of the problem. In recent years, Blochmann [106] and Sedgwick [107] have tried to salvage the gastraea theory by the idea that both mouth and anus are formed from a slit-like blastopore. They refer to Balfour's studies on *Peripatus* and to published data on *Aplysia* and other Gastropoda, and they consider the slit-like blastopore as the origin of both mouth and anus to be universal in all Metazoa. The major problems of the gastraea theory, however, cannot be solved in this way. According to Sedgwick's view, the radially symmetrical gastrulas of the Echinodermata, *Pilidium* and *Polygordius* should be regarded as larval forms with quite considerable secondary modification. Embryos of the Gastropoda, *Peripatus*, insects, and worms with slit-like blastopores represent, in this view, a primitive form of gastrula. Following the same reasoning, regular blastulas of the above-mentioned animals should be regarded as modified descendants of amphiblastulas of forms rich in yolk.

In the phylogeny of the anus, which is inadequately explained by the above theories, we see the same stages as in the formation of the mouth. In the lower Metazoa, we can find two or more apertures to eject the [waste] products of secretion, similar to the multiple incurrent pores of medusae. In some medusae of the Lafoeidae (e.g. *Aequorea, Tima*), the numerous outlet pores of the gastrovascular system occur on special pappillae; some Polyclada have similar excretory pores on different parts of the body. In *Cycloporus*, Lang [108] observed the excretion through such pores of drops of liquid with variously colored inclusions from the process of digestion. This discovery is especially important because the Polyclada have a unique excretory system, different from other [Turbellaria].

While part of the phagocytoblast was transformed into endoderm, in which the amoeboid cells gradually acquired an epithelial form, the rest of this primary tissue gave rise to mesoderm.

Initially, the latter was composed of separate motile cells which acted as phagocytes had behaved earlier. Similar situations are found in many coelenterates (e.g. in some corals, according to Kovalevsky and Marion, where wandering endoderm cells can appear on the scene), and in Echinodermata, worms, etc. Thus, there is something like "hemodynamics" in the development of mesoderm, for, in a variety of cases, it can develop through the migration of cells. While the entire mesoderm, or part of it, may retain its primitive state as phagocytoblasts, peculiar mesodermal sacs may be formed in many animals; because these sacs detach from the endoderm, their cell walls lose their phagocytic function to a greater or lesser degree. This may be seen in Ambulacralia and vertebrates. Such sacs have been searched for in vain in other animals with simpler forms of development, such as worms, mollusks and many arthropods; perhaps their entire mesoderm develops without the participation of true endodermal sacs, possibly because of the form of the ectoderm. In such cases, the mesoderm may later differentiate into somatopleura and splanchnopleura, without a sac-like stage. In Arthropoda with a significant amount of nutritive yolk, the mesoderm becomes active very early, its cells devouring and digesting yolk granules. This makes it clear why, until quite recently, mesoderm was often mistaken for endoderm. As a result of its phagocytic and kinetic functions, mesoderm plays an important role in embryonic development, and, sooner or later in evolution, it separated as an independent embryonic layer. In contradistinction to authors who propose the formation of mesoderm from sexual organs or muscles, I consider it to be part of the phagocytoblast, playing an important role in capturing and absorbing food.

As I have already elaborated in other works on the significance of mesodermal phagocytes in physiology and pathology, and also on the morphology of mesoderm, I merely refer to them here [109].

I should like to mention, by way of conclusion, that understanding the origin of multicellular animals is crucial to laying the foundations of comparative morphology. However, as long as the antiphylogenetic approach to the interpretation of issues concerning the embryonic layers persists, [the search for] this understanding will encounter major obstacles. That is why, in the continuing absence of positive data, I regard hypothetical explanations to be legitimate.

REFERENCES

[1] *Memorie per servire alla Storia di Polypi marini*, Napoli, 1785.

[2] Bidrag til Soedyrenes Natur-Historie, Bergen, 1829, V.1, translated in *Oken's Isis*, 1833, N 3, p. 221.

[3] Beskrivelser or Jagttagelser over nogle maerkelige eller nye i Havet ved den Bergenske Kyst levende Dyr, Bergen, 1835, summarized in *Archiv f. Naturgeschichte*, v. II, 1836, S.197.

[4] Zut Entwicklung d. Mollusken und Zoophyten. *Archiv f. Naturgesch.*, v. I, S. 406.

[5] Vorlaufige Miiheilung einiger unbekannter Structurverhaltnisse, bei Acalephen und Echinodormen. *Archiv f. Anat. u. Physiol.*, 1834, S.562, and Ueber die Acalephen des Rothen Meeres. *Abhandl. d. Akademie d. Wiss. zu Berlin aus d. Jahre 1835*, Berlin, 1837, S. 196.

[6] Ueber die Geschlechtsorgane der Medusa aurita. *Froriep's Notizen*, September 1836, N 1081 (v. L, N 3), SS. 33–35.

[7] Die Jungen der Medusa aurita. *Froriep's neue Notigen*, November, 1838, N 166 (v. VIII, N12), SS.177–180.

[8] K. Siebold. *Beitrage zur Naturgeschichte der wirbellossen Thiere*, 1839. I. Ueber Medusa aurita, S.1–35; and Sars. Ueber die Entwicklung d. Medusa aurita u. Cyanea capillata. *Archiv f.* Naturgesch., v. I, S. 9.

[9] *The Edinb. Philos. Journ.*, v. XXI, 1836.

[10] Ueber eine neue Art von nackten Armpolypen. *Oken's Isis*, 1833, N 3, S. 256.

[11] K. Svensk. Vetensk. Acad. Handl., 1835, S. 260; German translation in *Archiv f. Naturgesch.*, v. I, 1837, S. 249 and 321.

[12] "Salps are similar to medusae in that it is not the larva but its derivative that develops into the mature animal, not the individual but the method of generation that undergoes metamorphosis." (*Arch. f. Naturgesch..*, v.I, 1841, S. 29.)

[13] *Ueber d. Generationswechsel oder die Fortpflanzung u. Enrwicklung u. Entwicklung durch wechselnde Generationen. Uebers. v. Lorenzen*, Kopenhagen, 1842.

[14] See in *Archiv f. Naturg.*, V.II, 1843.

[15] Memoire sur le developpement des Meduses etc. *Annales des sciences naturelles*, Series III, V.IV, p.273.

[16] Ueber eine eigenthumliche Meduse des Mittelmeeres und ihren Jugendzustand. *Archiv f. Anat. u. Physiol.*, 1851, S.272, Fig. XI. Several years earlier J.Muller described young jelly-fish larvae from Niece (*Abhandl. d. Akademie d. Wiss. zu Berlin*, Jahrg. 1849, Fig. VII, Tabl. 9–11), which later he was inclined to consider young *Polyxenia*. They are, however, young *Aglaura* and identical with *Trachynema ciliatum Geg.*

[17] Ueber die Erzeugung v. Schnecken in Holothurien. *Archiv f. Anat. u. Physiol.*, 1852, S. 35.

[18] Ueber die fruhesten Entwicklungsstufen von Pelagia noctulica. *Ibid.*, 1855, S. 491, Tabl. XX.

[19] *Rare and Remarkable Animals of Scotland.* London, V.I, 1847, p.77 and fol., Fig. XII-XX.

[20] *Beobachtungen uber Anatomie u. Entwickling einiger wirbellosen Seethiere*, Berlin, 1851, S.25 and fol.

[21] *Zur Lehre vom Generationswechsel u. die Fortpflanzung bei Medusen und Polypen*, Wurzburg, 1854.

[22] Bibliography of these texts see in: Hinks. *A History of the British Hydroid Zoophytes*, V. I, London, 1868, pp. 327–330; and in: Hewkes. Bibliography to Accompany Selections from Embryological Monographs, *Bull. of the Mus. of Com. Zool. at Harvard College*, V. XI, N10. This bibliography does not contain many important texts, like just mentioned ones by Hinks, a well as some works by Bush, Leucart, F.E. Schultze and others.

[23] Untersuchung uber die Entwicklung d. Coelenteraten. *Gottinger Nachrichten*, 1868, N 7, S. 154.
[24] *Bull. de l'Academie Imp. des Sciences de St.-Peterbourgh*, V. VIII, 1865, p. 218.
[25] Nabludeniye nad razvitiem Coelenterata. Izv. O-va lubit. estestv., antropol. i etnogr. Separate brochure was published in 1873, then in 1874 – vol. X, N 2.
[26] Studien Uber die Entwicklung der Medusen u. Siphonophoren. *Zeitschr. f. wiss. Zool.*, V. XXIV, 1874, S.15 and foll., Fig.II-V.
[27] Die erste Entwicklung des Geryonideneies. *Jen. Zeitschr.*, vol. VII, 1873, S. 474, Tabl. XXIV, Fig.1.
[28] Studien uber die Entwicklung d.Medusen u.Siphonophoren. *Zeitschr. f. wiss. Zool.*, v.XXIV, 1874, S.17, Tabl.II, Fig.1.
[29] Das Problem der Befruchtung, *Jen. Zeitschr.*, v. XVIII, 1883, S. 286 and fol.
[30] *Metagenesis u. Hypogenesis von Aurelia aurita*, Jena, 1881, S. 10.
[31] Beitrage z. Entwicklungsmechanik d. Embryo. *Zeitschr. f. Biol.*, v. XXI, 1885, S. 423.
[32] *Zeitschr. f. wiss. Zool.*, V. XXXVI, 182, S. 436, Tabls. XXVIII, Dr. ## 1,2,4.
[33] *North American Acalephae*, Cambridge, 1865, p. 115, Dr. 172.
[34] Ilya i Ludmila Metchnikovy. Materialy k posnaniyu meduz i sifonofor. *Isv. ob-va lubit. estestv., antropol. i etnogr.* [t. VIII, vyp. 1], M., 1871, str. 60.
[35] *Untersuchungen uber die Organisation u. Entwicklung d. Medusen*, S. 87.
[36] *Das System der Medusen*, 1879, S. 207.
[37] *Contributions to the Nat. Hist. of the Unit. St. of America*, v. IV, p. 306. *North Amer. Acalephae*, p. 78.
[38] *System der Medusen*, S. 168, 186, 187.
[39] *Untersuchungen uber die Organisation u. Entwinklung d. Medusen*, S. 87.
[40] *Annals and magazine of Nat. Hist.*, IV, v. XLIX, 1874, p. 237.
[41] Spongicola fistularis. *Archiv f. Microsk. Anat.*, v. XIII, 1877, S. 795, Tabl. XLV-XLVII.
[42] *Monographue der Medusen*, p. II, 181, S. 130.
[43] *Annals and Madazine of Nat. Hist.*, V. XLIII, 1871, p. 79, Tabl. V, Dr. #2.
[44] *Ibid.*, p. 131.
[45] On Young Stages of Limnocodium and Geryonia. *Quart. Journal of Microsc. Science*, N.S., v. XXI, p. 194 and follow.
[46] Ueber die Erzeugung u. Holothurien. *Archiv f. Anat. u. Physiol.*, 1852, S. 34.
[47] *Zur Lehre vom Generationswechsel und d. Fortpflanzung bei Medusen u. Polypen*, 1854, S. 51 and follow., Tabl. II, Dr. # 17–23.
[48] *Archiv f. Anat. u. Physiol.*, 1851, S. 22 and follow., Tabl. XI.
[49] *Zeitschr. f. wiss. Zool.*, v. XXIV, 1874, p. 22 and follow., Tabl. III, IV.
[50] Ueber d. fruehesten Entwicklungsstufen d. Pelagia noctiluca. *Arch. f. Anat. u Physiol.*, 1855, p. 491, Tabl. XX.
[51] *Contributions to the Nat. Hist. of the Unit. St. of America*, v. III, 1860, Tabl. XII; 1862, p. 127.
[52] See his text on intestinal-cavitals quoted above, p. 7, Tabl. III.
[53] *Metagenesis u. Hypogenesis v. Aurelia aurita*, S. 28.
[54] *Cunina Kollikeri Archiv f. Naturg.*, 1861, S. 42, Tabl. IV.
[55] *Familie der Russelquallen*, 1865, S. 153 and the follow.
[56] *Zeitschr. f. wiss. Zool.*, v. XXIV, S. 27. Comp. to Metchnikoff's "Medusologische Mittheilungen". *Arbeiten d. zool. Inst. d. Univ.* zu Wien, v. VI, S. 251.
[57] *Ueber d. Cuninen-Knospenauhren*, Graz, 1875, S. 33.
[58] *System d. Medusen*, S. 313.
[59] *Die Russelquallen*, S. 156.
[60] Op. cit., S. 47.
[61] *Russequallen*, S. 129.
[62] Drawing of such an edging body can be found in my Russian-language book quoted above (Tabl. VI, Dr. # 3), as well as in *Medosologische Mittheilungen* (Dr. # 35).
[63] O. and P. Gertwig (*Organismus d. Medusen*, S. 17) describe one specimen of *C. lativentris (C. proboscidea)*, of 5 mm in diameter.
[64] *Ueber die Entwicklungsgeschichte der Thiere*, V. 1, 1828, Scholion V, S. 199 and follow.
[65] Herbert Spencer. *On Social Organism*
[66] "Istoriya razvitiya lancetnika (*Amphioxus lanceolatus*)," M.A. thesis, in Russian, 165, and

Entwicklungsgeschichte des Amphioxus lanceolatus. *Mem. Acad. Sciences de St.-Petersbourgh*, [v. XI, # 4], 1867.

[67] "Istoriya razvitiya *Sepiola*." M.A. thesis, in Russian, 1867. See summary by Klapared in *Archives des Sciences physiques et naturelles*, 1867. Compare to Grenaher's work published in *Zeitschr. f. wiss Zool.*, v. XXIV, 1874, on evolution of cephalopods written in his old manner, where embryonic layers were not mentioned at all.

[68] "Istoriya razvitiya *Nebalia*." Ph.D. thesis, in Russian, 1868.

[69] Entwicklungsgeschichtlige Beitraege, 1868. *Melanges biologiques de l'Acad. Sciences de St.-Petersbourg*. Embryonic leaves were strongly rejected by Ganin, in his Russian-language thesis (1867) on embryology of scorpio, and in *Zeitschr. f. wiss. Zool.*, v. XX, 1870, S. 518. When Kovalevsky (Embryol. Studien ueber Wuerner u. Arthoropoden, 1871, S. 5) claimed that Ganin "presented positive data that neural system emerges from upper cellular layer", this was based on misunderstanding because the above-mentioned researcher traced neural system of Scorpio directly from embryonic stripe and did not admit differentiation to cellular layers. It was another false move on Kovalevsky's part when he attributed to Ganin discovery of intestinal-glandular layer.

[70] See a summary in *Zhurnal Min. nar. prosv.*, April, 169, p. 179. I cannot agree with Kovalevsky when he (*Studien ueber Wuerner und Arthropoden*, S. 1) used this sentence of mine to conclude that I do not admit "any comparison between embryonic layers of vertebrates and invertebrates".

[71] Embryol. Studies an Wuerner u. Arthropode. *Mem. de l'Acad. Sciences de S.-Petersbourg*, v. XVI, # 12, 1871. Kovalevsky wrote on p. 5: "Embryonic layers which I accept, have only one thing in common with the ones described by Metchnikoff, that they are embryonic layers". Unfairness of this claim can be proved by comparison of our data; lately Kovalevsky himself studied scorpions and became convinced that embryonic layers described by me completely coincide with his description. If before he did not compare the lower layer of insects to the intestinal-glandular layer of vertebrates and, therefore, to the one of scorpion, he certainly has changed his mind now. I must Kovalevsky's incorrect interpretation of my ideas as ignoring my participation in studies of attribute embryonic layers.

[72] See his works on *Dicyemidae* in *Bull. de l' Acad. Sciences de Belgique*, 1876, and *Archives de Biol.*, v. III, 1882, p. 22, as well as works of Jullene, Goetthe and others.

[73] *Studies zur Gastraea-Theorie*, 1877, S. 258.

[74] *Vergleichende Anatomie des Nervensystems und Phylogenie der Mollusken*, 1877, S. 21.

[75] *A Manual of the Infusoria*, v. II, 1881–1882, p. 480.

[76] *Studien ueber die ersten Entwicklungsvorgaenge der Eizelle etc.*, 1876, S. 219.

[77] Bemerkungen zur Gastraea-Theoris. *Morphol. Jahrb.*, 1883, S. 124.

[78] Ueber *Palmellaceen* u. einige *Flagellaten*. *Archiv f. mikroskop. Anat.*, v. VIII, 1871, S. 429, Tabl. XXIV, Dr. # 32.

[79] *Der Organismus der Infusionsthiere*, V. III, Chapt. 1, 1878, Tabl. VII, Dr. ## 4,8.

[80] *A Manual of the Infusoria*, v. III, 1880–1882, Tabl. IV, Dr. #14.

[81] *Ibid.*, Tabl. VI, Dr. ## 2–5, 27–29.

[82] Comp. Buetschli's summary in Bronn's *Klassen und Ordnungen*. Protozoa, and his *Bemerkungen* zur Gastraea-Theorie, S. 423; further, Schtein's *Organismus*, v. III, Tabl. XVII, XVIII. I also observed first stages of division in *Volvox* and traced the formation of lamellar with its subsequent transformation in open bubbles.

[83] *Gastraea-Theorie*, S. 267, also S. 247, note; comp. the latest Haeckel's work: *Ursprung u. Entwicklung d. thierischen Gewebe*, 1884, S. 50.

[84] Der Organismus der Hydropolypen. *Jen Zeischr.*, v. XV, 1882.

[85] Haeckel also calls various larvae of medusae, two-layered planula, larva *Geryonia*, etc. simply gastrula not giving any reasons for this and citing no ground for its correspondence with his own definition of gastrula. See, his *System der Medusen*, SS. 119, 298, 348, etc.

[86] Notes on the Embryology and Classification of the Animal Kingdom. *Quart. Journ. of Microscop. Science*, N.S., v. XII, 1877, p. 403.

[87] On the Morphology a. Syst. Posit. of Spongida. *Quart. Journ. of Microscop. Science*, N.S., v. XIX, p. 103, *and Handb. d. vergl. Embryologie, uebers. v. Vetter*, v. I, 1880, S. 143.

[88] Zur Metamorphose von Oscarella lobularis. *Arbeiten s. zool. Inst. d. Univ. zu Wien*, v. VI, S.2, 1886,

S. 227. In this important work Geider pays too much attention to gastrulation in *Oscarella* and misses in his theoretical consideration modes of endoderm formation in other sponges.
[89] Bemerkungen zur Gastraea-Theorie. *Morphol. Janrb.*, v. IX, S. 415.
[90] Comp. to my "Vergleichend-embryologische Studien, 4", in *Zeitschr. f. wiss. Zool.*, v. XLII, 1885, S. 650.
[91] Die Metamorphose von Sycandra raphanus. *Zeitschr. f. wiss. Zool.*, v. XXI, 1878.
[92] Spongiologische Studien. *Zeitschr. f. wiss. Zool.*, v. XXXII, 1879. S. 374. Same ideas were expressed in 1877, in three Russian-language articles in Izv. Novoross. ob-va estestvoisp., v. V.
[93] *A Manual of the Infusoria*, v. I, 1880–1881, p. 353, Tabl. X, Dr. ## 20–30.
[94] *Zeitschr. f. wiss. Zool.*, v. XXX, 1878, S. 416, Tabl. 416, Dr. # 30.
[95] *Handb. d. vergl. Embryol.*, v. II, S. 308.
[96] *Zeitschr. f. wiss. Zool.*, v. XXXII, S. 382.
[97] Untersuchungen ueber die intracellulaere Verdauung. *Arbeiten des zool. Inst. d. Univ. zu Wien*, v. V, 1883, S. 142.
[98] *Zeitschr. f. wiss. Zool.*, v. XXXII, S. 374.
[99] Oscarella lobularis (see, note 88), S. 56.
[100] *Handb. d. vergl. Embryol.*, v. I, 1880, S. 172.
[101] *Abhandlungen z. Entwicklungsheschichte d. Thiere*, v. II, 1884, S. 184 and foll.
[102] Sur le development des Turbellaries. *Zool. Anz.*, v. VIII, 1885, N 194, p. 269.
[103] *Ibid.*, p. 272.
[104] See, his dissertation, in Russian, on *Dinophilus*, Odessa, 1886, p. 47.
[105] *Zeitschr. f. wiss. Zool.*, v. XXXVII, 1882, S. 286.
[106] Beitrauage z. Kenntn. d. Entwicklung d. Gasteropoden. *Zeitschr. wiss. Zool.*, v. XXXVIII, 1883, S. 406.
[107] On the Origin of Metameric Segmentation, etc. *Studies from the Morphol. Laboratory in the Univ. of Cambridge*, v. II, 1884, p. 1, p. 82.
[108] *Die Polycladen (Faunna und Flora d. Golfes von Neapel)*, 1884, S. 158.
[109] Vergleichend-embryologische Studien, 3–5. *Zeitschr. f. wiss. Zool.*, v. XXXVII, 1882, S. 286, and v. XLII, 1885, S. 648. – Untersuch. ueber die intracellulaere Verdauung bei Wirbellosen, *Arbeiten des zool. Inst. d. Univ. zu Wien*, v. V, 1883, S. 147. – Untersuch. ueber die mesodermalen Phagocyten einiger Wirbelthiere, *Biol. Centralbl.*, v. III, 1883, S. 560. – Ueber eine Sprosspilzkrankheit der Daphnien. *Virchow's Archiv*, v. XCVI, 1884, S. 177. – Ueber die Beziehung der Phagocyten zu Milzbrandbacillen. *Ibid.*, v. XCVII, 1884, S. 502.

NOTES

1 We have used "Medusae" as the general term, restricting "jellyfish" to the large Scyphomedusae (= Scyphozoa).

2 A family of freshwater sponges.

3 The central theme of the phagocytosis theory formally developed between 1882 and 1891.

4 See M.T. Ghiselin and C. Groeben "Elias Metschnikoff, Anton Dohrn and the Metazoan Common Ancestor". *J. Hist. Biol.* 30:211–228, 1997.

5 *Cumina* seems to be a misprint for *Cunina* Eschscholtz, 1829.

6 The generic name *Sertularia* was used by Linnaeus (1758) for a hydroid; Péron & Lesueur (1810) named a medusa *Obelia* before it was realized that one developed into the other. The Code of Zoological Nomenclature insists on using the oldest valid names, so, when *Sertularia* was split into several genera, one of them became *Obelia,* and the species originally known as *Sertularia geniculata* became *Obelia geniculata*, implying a medusa with knees.

7 = *Aurelia aurita.*

8 Only two are described here.

9 *i.e.* in terms of contemporary morphology.

10 near Nice.

11 More correctly, Oceaneidae and Campanularidae. We have retained Metchnikoff's zoological names throughout, although not always current.

[12] Invagination is "pushing in" rather than "sticking out".
[13] Literally "lacking a velum". Acraspeda = Acraspedota = Scyphomedusae.
[14] Ambulacralia = Echinodermata + Enteropneusta.
[15] Emendation of Gertwig, here and subsequently.
[16] Without a hydroid phase.
[17] With alternation of generations.
[18] Metchnikoff, like most of his contemporaries, used an initial capital for specific names derived from personal names, but today this species would be *Microtoma annae*.
[19] Multipolar ingression.
[20] Geryoniidae.
[21] Symmetrically arranged parts, such as tentacles.
[22] An early recognition (perhaps the first) that classification of hydroids and their corresponding medusae are incompatible, which is difficult to explain in terms of lineal evolution.
[23] Aequoreidae.
[24] This assumes that the medusoid and hydroid generations have always evolved as part of the same life-history, but the continuing failure to find compatible classifications for the two phases is consistent with Williamson's view that they were originally separate groups that came together as the result of several hybridizations.
[25] *Eucope* Gegenbauer, 1856, is the original name of this genus, and it is still in use. *Eucopia* Boeck, 1861, was proposed as a substitute, but this generic name was already in use for a crustacean (*Eucopia* Dana, 1852) and is therefore invalid. *Eucopium* Haeckel, 1879, was proposed as a genus distinct from *Eucope*.
[26] short life-cycle or direct development.
[27] alternation of generations.
[28] Metchnikoff knew that these larvae had cilia, and he drew them (*fide* Hyman).
[29] i.e. regarded as invalid.
[30] The manubrium is very long and protruding in this genus, hence "outer".
[31] = Cuninidae.
[32] This ignores Haeckel's suggestion that two species are involved.
[33] Families (with corresponding orders Orthonectida and Dicyemida) of the phylum Mesozoa.
[34] perhaps "cytopharynx".
[35] Aphididae.
[36] for *Protospongia*. It eats minute food particles, but the Volvocales are holophytic.
[37] This must refer to development from the 2-dimensional, circular *Gonium*-like stage to the 3-dimensional, spherical *Volvox*, with non-reproductive cells at the "anterior" pole.
[38] An archiblastula consists of one type of cells, an amphiblastula of two types.
[39] Phylum Placozoa, previously regarded as an aberrant mesozoan or a modified hydroid planula.
[40] presumably, tubular or trumpet-shaped sponges.
[41] multinucleate mass, part of an amoeboid organism.

CHAPTER 7

THE STRUGGLE FOR EXISTENCE BETWEEN PARTS OF THE ANIMAL ORGANISM (1892)

We are all aware of the terrible struggle with the natural catastrophe that has so unexpectedly befallen us.[1] At this time when we all unite to overcome this disaster, it seems particularly interesting to make a thorough study of the organism's struggle for existence. On the one hand, it is very important to study the conditions that provoke the intense struggle, since only in that way can we find means for its prevention, but, on the other hand, it is of no less interest to analyse the phenomena that occur in the struggling animal itself.

I

A significant part of the struggle for existence, in life in general and in the development of organisms in particular, was realised long ago. Although Lamarck, who put forward his theory of the origin of species at the beginning of this century, did not emphasise this aspect, he did contend that evolutionary changes took place in response to the urgent needs of the organism. In his much quoted example of the giraffe, Lamarck sought to explain the development of the long neck and long legs as resulting from its predecessors' strivings to tear down leaves from tall trees. The urgent need to find food in the African climate, with its frequent droughts that burn all pasture, directed this striving. In this example, as in so many others discussed by Lamarck, an initial stimulus generates an active striving for its resolution. Prolonged exertion toward this goal promotes the active development and appearance of new organs. Accordingly, this should result in significant changes in the organism and lead to the creation of new species.

This aspect of Lamackian theory was enthusiastically echoed in the philosophy of Schopenhauer. By treating will as the supreme worldly principle, the Frankfurtian thinker sought to explain changes in the organism as an activity of will directed to the satisfaction of its needs. So, under the influence of will, the neck and legs of the giraffe became longer, bats developed wing membranes, and, in general, many organs were formed and improved.

Together with this principle of activity, Lamarck implied the opposite effect, determining the underdevelopment and disappearance of organs. While active utilisation might lead to the extension of neck, legs, and other parts of the body, under-utilisation diminishes them. In this way, the eyes of many subterranean animals have atrophied, precisely because of their prolonged disuse, *i.e.*, in a purely passive way.

Although Darwin quite clearly demonstrated the impossibility of applying Lamarckian principles to the explanation of every phenomenon related to the origin and transformation of species, he still regarded utilisation and non-utilisation of organs as important, and considered them as instrumental in the development of organic forms. Together with these factors, he also emphasised another – natural selection, *i.e.*, the preservation of individuals with the best adapted innate characteristics. This selection is directly related to the struggle for existence: the more powerful the selection, the stronger the struggle for existence (as well as the need that provokes it).

Regardless of the stand-point from which we view the origin of species, we see variability as a response to need everywhere, and this variability manifests itself in the eternal, and sometimes extreme, struggle for existence. It is not possible here to detail Darwinian theory, but we should note that he always drew a clear line between two major factors – natural selection and deselection in the struggle for existence. Natural selection preserves changes that are beneficial to the organism for any reason, while the Lamarckian principle can operate without reference to individual improvement.

That is why Darwin, applying his theory to particular cases, always raised the question of benefit as a necessary indication of natural selection. Thus, referring to the blindness of subterranean animals, he admitted that eye-loss may be beneficial in cases where frequent inflammation of the eyes results from the subterranean way of life. Organs which become useless because of the absence of light, thus turn out to be harmful because of their proneness to inflammation, and thereby dangerous to the entire organism. Removal of such an organ would thus be determined, to a considerable degree, by active natural selection. So, in general, "as it is difficult to imagine that eyes, though useless, could be in any way injurious to animals living in darkness, I attribute their loss wholly to disuse."[2]

As is well known, scientists fall into two camps on the explanation of facts regarding the origin of species. The first camp puts forward the Darwinian principle of selection, while the other remains faithful to the Lamarckian theory of the immediate influence of use and disuse. Everyone, nevertheless, recognises the involvement of both factors in the origin of species, although to different degrees. In order to form some opinion on the part played by selection in each of them, and, in general, to penetrate deeper into the field of variability of organic forms, a closer analysis of what occurs in the depths of the organism seems necessary.

II

At the beginning of the 80s a young and very talented German scientist, Wilhelm Roux, made an interesting attempt to determine the conditions under which organs change, either by enlargement on the one hand or reduction on the other.

Since increased or decreased use of organs seems to be associated with enhanced or diminished blood supply, Roux considered this hypothesis first. It is easy to show, however, that this factor is of only secondary importance in modifying organs. It has long been known that an artificially enhanced blood supply to a rabbit's ear, lasting for up to a week, does not lead to an increase in the size of that organ. Roux knew a doctor whose blood vessels on the distal joints of both little fingers were significantly enlarged for seven years, but, in spite of this, the skin looked absolutely normal and the joints themselves were not enlarged. On the other hand, in cases of abnormal growth of certain parts of the body, the blood vessels do not become noticeably dilated. Prolonged compression of the arteries supplying such enlarged parts of the body never seems to lead to their shrinkage.

It is clear, therefore, that increased blood flow is not of crucial significance in the case of abnormal growth of certain parts of the body. This phenomenon occurs as a result of heightened activity of the smallest components of certain organs, the so-called cells. If we consider the case of enlargement of a muscle due to increased exercise, the cause is not the augmented blood supply as such but the activity of the muscle's components, *i.e.*, the physiological function of cells that make up this organ.

The change of opinion on the overdevelopment of an organ is comparable to that on the cause of inflammation. The increased blood flow, which is the most noticeable feature, led to the long-lasting assumption that the cause of inflammation was precisely this rush of blood. It later became evident, however, that a rush of blood could last a long time without any inflammation, and was not followed by any sweating. The determinant of the process therefore appeared not to be an overfilling of the inflamed organ by blood, but the particular activity of the cellular components of the inflamed organ.

Analysis of the major processes in the mutability of organisms therefore led to the need to take a closer look at the functions of cells, these microscopical components of the vast majority of living creatures.

Trying to solve this problem, Roux came to the conclusion that the innumerable multitude of cells that compose the organism find themselves in a constant struggle with each other for food and space. Just as plants sown too thickly in a confined space harm each other by their close proximity, neighbouring cells fight for the nutrients required for their existence. Those plants and cells which, because of individual characteristics, can capture larger quantities of nutritious material, defeat their weaker partners. As a result of this inner struggle between components, the organism becomes reinforced, and only the strongest cells remain. The more protracted the concurrence, the stronger the remaining cells become. An individual composed of such strong cells is definitely better prepared to struggle for its existence than individuals of the same genus with weaker cells.

Roux gives only an indirect answer to the question of the distinctive features of successful cells. According to his opinion, the most responsive cells are those most capable of attracting nutritious juices and, therefore, of defeating their less responsive neighbours. The very reinforcement of sensitivity, which is a condition of victory and survival, leads to an intensification of various specific functions of the organism.Thus a struggle for existence between the smallest components of an organism is a necessary condition for this organism to be successful in the more obvious struggle with other individuals; this situation, therefore, plays an important part in the transformation of organic forms themselves.

III

Roux's proposal is quite sound in the following respect: it shows that there is an urgent need to undertake a careful analysis of the most intricate processes occurring in cells whenever we try to understand the facts of variability of organs and organisms. Strictly speaking, this conclusion has been evident, in one sense, for a long time. If understanding the functions of some organs seems impossible without analysis of the functions of their microscopic components, the same should be presumed when the increased or decreased development of organs, their variability as it were, is considered.

During the ten years since the publication of Roux's manuscript, the scientific investigation of the microscopical components of organisms has made significant progress; as a result, the problem of the inner struggle of cells within the organism must be subjected to substantial revision. Let us use the latest discoveries in order to facilitate our understanding of the phenomena of variability in the organic world.

Since we are powerless to analyse changes that have affected organic forms over many generations, we must turn to phenomena that occur before our eyes, and which are followed by significant changes in outer form and inner organisation. We do not know what changes led to the appearance of the first flies and butterflies, starfish and frogs, as well as many other animals, but we may certainly follow the complicated transformations by which these creatures develop today. Flies and butterflies, before they appear in their final shape, live for a long time as slow-moving wingless larvae and caterpillars; ponderous starfish, [seemingly] overfilled with mercury, first appear as transparent many-armed larvae that swim with ease at the sea-surface; frogs, as is well known, originate from tadpoles, which, because of their tails and absence of legs, resemble fish more than reptiles. Metamorphosis in all these animals, like the creation of new species, involves the development of new organs and the disappearance of old ones, the only difference is that such contemporary transformations happen in just a few days, whereas transformations of species require an immeasurably longer time. But these two processes are of comparable significance: the struggle for existence between cells of an organism plays an important role, and the formation of new organs is possible only as a result of the defeat of weak cells by strong ones. Numerous recent researches reveal the mechanism of this struggle, which is, in fact, considerably less complicated than might have been supposed under Roux's theory.

There are many cells in the organs of the overwhelming majority of animals which are quite similar to the most primitive "rootlegs" , amoeba, and which, like it, can eat different solid objects, including various types of cells. These amoeboid devouring cells, or, as they may be more briefly called, phagocytes, eat all parts of the organism which have become weak for any reason, while ignoring parts fully capable of living. This is precisely what happens in the body of a starfish larva when it undergoes transformation into a juvenile starfish. While early rudiments, composed of partially isolated cells, grow normally and are not bothered by phagocytes, other larval organs are attacked and, in a short time, are either eaten completely or small residual fragments are left. The struggle which takes place in larval cells is mainly between different parts of various organs and the phagocytes. Young cells survive the struggle, and old cells usually fall victim to phagocytes. The apparent result of this struggle is extremely rapid changes in the organs to transform the larva into a starfish, which bears not even the slightest resemblance to its predecessor.[3] The changes affect not only the outer form of the animal but its entire way of life.

Phenomena described with respect to the starfish may seve as a prototype for the dramatic transformations in animals in general. Especially instructive here is the story of the common fly, so splendidly narrated by my famous compatriot A.O. Kovalevsky and the Dutch scientist van Riss. The majority of organs and cells which compose these organs in vermicular fly larvae fall victim to phagocytes which penetrate their muscle fibers, salivary glands and other organs, devouring all their contents and changing a significant part of the body into a thick mass resembling cream or puss. At the same time, many young cells, spared by phagocytes, begin to transform themselves into new muscles and new glands, so that the organism is rebuilt, almost from scratch, and completely regenerated. New organs developed in this way, in spite of the proximity of voracious phagocytes, are distinguished by quite amazing stability and power. They represent the lucky chosen ones in this natural selection, which takes place in the depths of the body. It is not surprising, therefore, that flies are, in general, more tenacious of life than their relatives, such as mosquitoes, whose tissues do not undergo such a complete renewal at metamorphosis.

Evidence that this struggle between ordinary cells and phagocytes is not merely an isolated case, but has universal significance and is inherent in both lower and higher vertebrate animals, is seen, with most instructive details, in the metamorphosis of frogs.

At the very time when a tadpole is still swimming, in good condition, with the help of its broad tail, its tissues already show signs of profound change. Many muscles become pervaded by phagocytes, which gradually consume all the contractile mass and transform the muscle into a conglomerate of active cells overflowing with the remains of phagocytized fibers. This devouring then spreads to nerves, skin and skeleton (spinal cord), so that the entire content of the tail is finally transformed into a kind of thin gruel, consisting of phagocytes overfilled with food. The latter, with their amoeboid movements, move slowly along the tail, and finally take the remains of the devoured fibers into the general body cavity of the young frog. So, the tail virtually disappears. The disappearance or atrophy of the tail thus occurs because of phagocytic activity, which is resisted by only the most vigorous elements of the organism.

Besides the tail, certain other parts of the tadpole, like branchiae, are subjected to similar consumption, whereas the legs, lungs and many other organs continue their further development without interruption.[4]

We thus see that an intense struggle between cells plays the key role in this accelerated type of transformation, which [over many generations] leads to the extermination of certain animals and the appearance of others, and this struggle may be reduced primarily to resistance to the devouring activity of phagocytes. Quite possibly similar processes have taken place in a much longer sequence of events, which determined the origin of species.

IV

In evaluating the action of phagocytes, there is a certain trend towards interpreting this activity simply as a means of getting rid of cells and organs that have become "unnecessary" for the organism; thus there is a tendency to introduce teleology into the explanation of events. In order to understand the phenomenon, however, it seems important to consider other cases where comparable activity of these voracious cells reveals itself. Such examples are everywhere. At any time there are numerous phagocytes in the spleen and liver of man and other higher animals, and such phagocytes are full of different particles, especially red and white blood corpuscles. Inside the phagocytes, these cells undergo a series of changes resulting in their dissolution and digestion by the phagocytes. In this example we see a continuous struggle for existence between cells, and this is not merely limited to the particular period of metamorphosis. The organs just mentioned, particularly the spleen, serve, so to speak, as the animal's blood purifiers, removing weakened blood corpuscles whose place is immediately taken by newly formed, stronger cells. This activity results in the regeneration of the blood, this most important tissue of humans and other vertebrates.

But it is not only blood that has a guard to protect it from weak cells. There are a number of such guards in humans and the majority of animals, guards in the form of different types of phagocytes. Of the active organs, the most important role is played by the relatively complex, striated muscle fibers. In these fibers, one can distinguish a contractile substance, consisting of very small prisms, and an interstitial plasma with numerous inclusions such as are found in any true cell. The contractile substance plays its role in movement, whereas the formless, interstitial plasma, with all its inclusions, is the guardian of the actual contractile muscles. If the latter, for any reason, become insufficiently strong, then the interstitial plasma consumes them, and thus allows new muscle fibers to grow. Each striated fiber, therefore, has its phagocytes, which remain close to the contractile substance itself. Because of its voracity, such a muscle phagocyte grows bigger and bigger, and this increase also applies to both the number and size of the inclusions within the phagocyte.

Atrophy of muscle fibers in the tail of a tadpole during metamorphosis to an adult frog occurs precisely by means of such muscle phagocytes. The contractile element becomes surrounded by formless interstitial plasma which then separates into a number of phagocytes, and these move deeper into the body cavity and acquire the form of mobile lymph cells.

The same phenomenon also takes place in damaged muscle fibers that have lost the capacity to contract. According to Dr. Sudakevitch, striated muscle fibers in contact with or impregnated by freshly prepared strychnine provide a very interesting and quite distinct example of the action of muscle phagocytes. Under the influence of this irritant, the formless muscle plasma thickens, its inclusions increase both in volume and in number, and the resulting phagocytes digest the contractile elements and counteract the strychnine. The contractile cells defend themselves by extruding so-called solid capsules, but these often perish like parasites under the attack of phagocytes. Some phagocytes, composed of a significant mass of plasma with numerous inclusions, form huge and often very complex cells known as "giant cells" or simply "giants".

Similar phagocytic activity also takes place in striated muscle during certain illnesses, so-called muscular atrophies, which are associated with the loss of contractile elements. The phenomena of the enlargement and division of cells, together with the formation of giants (which, incidentally, was described long ago), clearly indicate phagocytic activity.

Neural tissues also have their own phagocytes, guardians of the normal activity of nerve cells and fibers. The clearest evidence of this is provided by tissues which have been injured in some way. In damaged parts of the tissue, the membrane expands considerably due to increase in the number and size of its cells, and this results in the appearance of giant cells, very similar to those seen in muscles. Such neural phagocytes consume parts of the damaged nerve tissue, operating in the same way as muscle phagocytes. The latest studies of von Bugner leave not the slightest doubt on the correctness of this interpretation of these phenomena.

The peculiar tissue known as neuroglia has many star-shaped cells and probably takes the place of phagocytes in neural centres, protecting them from attack and simultaneously destroying weak cells in these essential organs. Further research is required, but the very fact that any loss of neural tissue is followed by local spreading of neurologlia clearly indicates an important role in this field.

Similar phagocytic phenomena may be observed in adipose tissue, followed by the spreading of surrounding plasma and inclusions. And it is especially important to note that these activities take place in adipose tissue in times of starvation, which shows the tremendous strength and endurance of phagocytes.

Prof. Podvysotskij has observed and described hepatic phagocytes that devour parts of dying tissue.

Even such a dense tissue as bone has its own phagocytes, appearing in the form of so-called bone-breakers, osteoclasts, *i.e.*, giant cells consisting of substantial amounts of plasma and numerous [small] cells.

Apart from this massive variety of specific phagocytes, there are many phagocytes of a so-called general nature, namely white blood corpuscles, which swim in the blood and make their way to any place where phagocytic activity is required. In many cases, when specific phagocytes are insufficient in strength or number, white blood corpuscles appear on the battle-field to reinforce phagocytic defence.

We thus see that there is a kind of state within a state in animals, a complex conglomerate of different cells. Numerous phagocytes, dispersed everywhere, maintain a considerable degree of independence while directing their activity to exterminating every weak member of the colony. If cells destroyed in this way can easily be replaced by new ones, then the result of such a conflict should be a considerable strengthening of the entire organism. We have already seen that the body of a fly renewed by the phagocyte's fight becomes much stronger after the war is over. Comparable examples occur in human life. We know of cases when, after a prolonged illness (especially after typhoid fever), the organism appears to be completely renewed, and the recovered patient becomes stronger and healthier than before the illness. What is happening in this case is a peculiar transformation similar to that observed in flies, frogs, and many other animals.

Such a strengthening after illness can be explained as resulting from the war between cells and phagocytes. Phagocytes stimulated by the fever which, in the case of typhoid, is prolonged, intensify their activity, while other tissues, like muscles and nerves, by contrast become significantly weaker. Phagocytes attack such tissues, and a brutal fight begins. As shown by Prof. L. Popov, white blood corpuscles can penetrate cells, and this leads, quite naturally, to the destruction of the latter. During typhoid fever, a number of phagocytic processes take place in muscles, including the devouring of many tissues. If, during this fight, phagocytes destroy vitally important elements, which either cannot be restored or there is not sufficient time for their regeneration, then the organism perishes completely, together with its phagocytes. If, by contrast, these watchmen remove all that are weak but do not undermine the integrity of the organism, and if the destroyed cells are replaced by new, fresher, stronger ones, then such a struggle results in a beneficial renewal.

How do phagocytes and their enemies act in this struggle for existence? For a long time the behaviour of mobile phagocytes (white blood corpuscles) was regarded as bizarre or even mystical, particularly because of their ability to move directly to often remote places to find their prey. Numerous recent investigations, however, have shown that phagocytes are capable of such discernment by so-called [chemotaxy]. They can determine the composition of the surrounding liquid, and, by assessing its quality, they are able to advance or take avoiding action. Phagocytes use this perceptivity not only in guiding their paths but also in selecting bodies to consume. It is known that substances emitted from damaged tissues attract phagocytes, which capture the dead cells greedily. On the other hand, it is not yet known why these phagocytes, in the proximity of healthy cells and in the absence of a definitive defence mechanism, do not attack these [healthy] cells. Perhaps the latter excrete special substances that repel phagocytes and hamper their activity. Only when the cell is strong and healthy does it secrete such a substance and the phagocyte leaves it untouched, but immediately it ceases to surround itself by this defensive substance, when sick or weakened, a phagocyte attacks and eats it. In this or some other way, cells have to protect themselves against attack by phagocytes and struggle for their own existence. They can survive and produce new tissues only if they win this struggle.

Such a processes of struggle, so widespread during metamorphosis and also during the disappearance of certain parts [of the body] and the creation of new ones, must take place both in the life of an individual and in changes of organisation which lead to the appearance of new species.

Let us apply the information set out here to the problem posed at the beginning of this paper, namely the disappearance of organs due to their disuse. The inactive tissues and cells of such organs may lose their capability to defend themselves against attack by phagocytes and therefore become susceptible to being eaten.[5] The eye of an animal that has moved to a dark cave is not stimulated by light and will consist of cells of weakened general tone. In such circumstances, defensive secretions and any other defence mechanisms will naturally be reduced or weakened. This usually results in such cells being eaten by phagocytes, and in the complete or partial atrophy of the targeted organ. The same can happen to an inactive muscle, nerve, or any other organ.

On the other hand, the appearance of a new part under the influence of active usage only can be related to a broader mechanism of phagocytic defence. One should not fail to take into account that phagocytic activity also takes place in the fetus, where it leads to the extermination of young cells and their rudiments. Whenever these cells cannot defend themselves against phagocytes, these devouring cells also eat the rudiments, in spite of their youth.

The breadth of phagocytic activity reveals a complex system of regulators that protect the strength of the organism and participate both in the destruction of organs and the formation of new ones, *i.e.* the major phenomena of mutability. Could a closer look at these aspects of the struggle within an organism lead to a solution of general questions on the mechanism of the variability of species, and also the role of natural selection and the question of immediate adaptability, in the context of Lamarckian theory?

Since an indication of selection is always the usefulness of certain characteristics, then we must ask whether the phagocyte system is generally useful to the organism.

Phagocytes, which may be considered very useful from the perspective of an individual's interests, might nevertheless do significant harm. Consider the example of a weakened tissue that is still functional and even capable of restoration to full strength: such an organ might be destroyed irrevocably by phagocytes. Such cases are probable relatively common in diseases of the central nervous system: nerve cells are easily poisoned or damaged but are unable to regenerate themselves. An attack on such cells by phagocytes can cause debilitating or fatal illnesses.

Such cases of phagocytes harmful to the organism may still contribute to the well-being of the genus, since they provide a mechanism of selection between individuals. They help to remove the weakest and preserve the strongest (or, more correctly, the removal of weak individuals leaves the strong ones). There is nothing teleological or envisaged in advance with respect to this advantage. It is easy to imagine a case in which the destruction of vital parts acquires epidemic proportions, and, as a result of which, the entire race or species may perish. We are unable, however, to provide an actual example of this phenomenon because our information is understandably incomplete.

On the other hand, it is quite evident that since the phagocyte system has proved so universally useful, it can be regarded as a result of natural selection. Even the destruction of inactive parts has its useful side: phagocytes pursuing such activities find food and have an opportunity to improve their effectiveness.

It is easy to assume, therefore, that variability, regarded as a general principle by Lamarck, is a necessary part of the wider aspects of selection, rather than opposed to it.

The phagocyte system is generally very useful , and because it works for the welfare of the organism is itself open to improvement by natural selection, but it has not yet reached its full potential. It may achieve even greater success when stimulated artificially. From the earliest times when humans became aware of the nature and peculiarities of phagocytic activity, they have been able to influence it. Analysis of the factors that strengthen or weaken phagocytes should lead to their manipulation to increase the potential benefits and decrease harmful effects. It is known that warmth and the vital secretions of certain microorganisms stimulate the activity of phagocytes, while cold, quinine, abrin, and some other substances suppress it. Science uses all this information and can add to it to improve on natural selection.

It becomes clear from this example, and from many others, that human consciousness, guided by positive science, acts like a prism, refracting and transforming natural selection to produce artificial selection.

NOTES

1 Metchnikoff refers here to one of the severe famines which were so frequent in Russia at that time.

2 *On the Origin of Species*. A facsimile. Cambridge: Harvard University Press. 1964. Ch.V, p. 137

3 Metchnikoff's account of starfish metamorphosis is rather misleading. Phagocytized larval cells are mostly replaced as the growing juvenile migrates through the larval body, and the larval organs are not transformed into those of the juvenile. In one species, *Luidia sarsi,* the swimming larva and crawling juvenile can co-exist for three months after their separation.

4 An explanation in terms of larval transfer is that phagocytes of the juvenile frog do not attack healthy frog cells but do attack the alien tadpole cells, originally acquired from another species.

5 Metchnikoff was to develop an entire medical theory based on the bodily economy of phagocyte function which continuously sought optimal expressions and thereby optimal health (e.g. *The Prolongation of Life. Optimistic Studies.* London: G.P. Putnam's Sons, 1910) For summary accounts see A.I. Tauber and L. Chernyak. "Metchnikoff and a Theory of Medicine." *J. Roy. Soc. Med.* 82:699–701, 1989 and A.I. Tauber, "The Immunological Self: A Centenary Perspective." *Persp. Biol. Med.* 35:74–86, 1991. For a review of Metchnikoff's relevance to late twentieth century immunology see A.I. Tauber, "Metchnikoff, the Modern Immunologist". *J. Leuk. Biol.* 47:561–67, 1990.

INDEX OF AUTHORS

INDEX OF ORGANISMS

37. H. von Helmholtz: *Epistemological Writings.* The Paul Hertz / Moritz Schlick Centenary Edition of 1921. Translated from German by M.F. Lowe. Edited with an Introduction and Bibliography by R.S. Cohen and Y. Elkana. [Synthese Library 79] 1977
ISBN 90-277-0290-X; Pb 90-277-0582-8
38. R.M. Martin: *Pragmatics, Truth and Language.* 1979
ISBN 90-277-0992-0; Pb 90-277-0993-9
39. R.S. Cohen, P.K. Feyerabend and M.W. Wartofsky (eds.): *Essays in Memory of Imre Lakatos.* [Synthese Library 99] 1976 ISBN 90-277-0654-9; Pb 90-277-0655-7
40. Not published.
41. Not published.
42. H.R. Maturana and F.J. Varela: *Autopoiesis and Cognition.* The Realization of the Living. With a Preface to "Autopoiesis' by S. Beer. 1980 ISBN 90-277-1015-5; Pb 90-277-1016-3
43. A. Kasher (ed.): *Language in Focus: Foundations, Methods and Systems.* Essays in Memory of Yehoshua Bar-Hillel. [Synthese Library 89] 1976
ISBN 90-277-0644-1; Pb 90-277-0645-X
44. T.D. Thao: *Investigations into the Origin of Language and Consciousness.* 1984
ISBN 90-277-0827-4
45. F.G.-I. Nagasaka (ed.): *Japanese Studies in the Philosophy of Science.* 1997
ISBN 0-7923-4781-1
46. P.L. Kapitza: *Experiment, Theory, Practice.* Articles and Addresses. Edited by R.S. Cohen. 1980 ISBN 90-277-1061-9; Pb 90-277-1062-7
47. M.L. Dalla Chiara (ed.): *Italian Studies in the Philosophy of Science.* 1981
ISBN 90-277-0735-9; Pb 90-277-1073-2
48. M.W. Wartofsky: *Models.* Representation and the Scientific Understanding. [Synthese Library 129] 1979 ISBN 90-277-0736-7; Pb 90-277-0947-5
49. T.D. Thao: *Phenomenology and Dialectical Materialism.* Edited by R.S. Cohen. 1986
ISBN 90-277-0737-5
50. Y. Fried and J. Agassi: *Paranoia.* A Study in Diagnosis. [Synthese Library 102] 1976
ISBN 90-277-0704-9; Pb 90-277-0705-7
51. K.H. Wolff: *Surrender and Cath.* Experience and Inquiry Today. [Synthese Library 105] 1976
ISBN 90-277-0758-8; Pb 90-277-0765-0
52. K. Kosík: *Dialectics of the Concrete.* A Study on Problems of Man and World. 1976
ISBN 90-277-0761-8; Pb 90-277-0764-2
53. N. Goodman: *The Structure of Appearance.* [Synthese Library 107] 1977
ISBN 90-277-0773-1; Pb 90-277-0774-X
54. H.A. Simon: *Models of Discovery* and Other Topics in the Methods of Science. [Synthese Library 114] 1977 ISBN 90-277-0812-6; Pb 90-277-0858-4
55. M. Lazerowitz: *The Language of Philosophy.* Freud and Wittgenstein. [Synthese Library 117] 1977 ISBN 90-277-0826-6; Pb 90-277-0862-2
56. T. Nickles (ed.): *Scientific Discovery, Logic, and Rationality.* 1980
ISBN 90-277-1069-4; Pb 90-277-1070-8
57. J. Margolis: *Persons and Mind.* The Prospects of Nonreductive Materialism. [Synthese Library 121] 1978 ISBN 90-277-0854-1; Pb 90-277-0863-0
58. G. Radnitzky and G. Andersson (eds.): *Progress and Rationality in Science.* [Synthese Library 125] 1978 ISBN 90-277-0921-1; Pb 90-277-0922-X
59. G. Radnitzky and G. Andersson (eds.): *The Structure and Development of Science.* [Synthese Library 136] 1979 ISBN 90-277-0994-7; Pb 90-277-0995-5

60. T. Nickles (ed.): *Scientific Discovery.* Case Studies. 1980
ISBN 90-277-1092-9; Pb 90-277-1093-7
61. M.A. Finocchiaro: *Galileo and the Art of Reasoning.* Rhetorical Foundation of Logic and Scientific Method. 1980 ISBN 90-277-1094-5; Pb 90-277-1095-3
62. W.A. Wallace: *Prelude to Galileo.* Essays on Medieval and 16th-Century Sources of Galileo's Thought. 1981 ISBN 90-277-1215-8; Pb 90-277-1216-6
63. F. Rapp: *Analytical Philosophy of Technology.* Translated from German. 1981
ISBN 90-277-1221-2; Pb 90-277-1222-0
64. R.S. Cohen and M.W. Wartofsky (eds.): *Hegel and the Sciences.* 1984 ISBN 90-277-0726-X
65. J. Agassi: *Science and Society.* Studies in the Sociology of Science. 1981
ISBN 90-277-1244-1; Pb 90-277-1245-X
66. L. Tondl: *Problems of Semantics.* A Contribution to the Analysis of the Language of Science. Translated from Czech. 1981 ISBN 90-277-0148-2; Pb 90-277-0316-7
67. J. Agassi and R.S. Cohen (eds.): *Scientific Philosophy Today.* Essays in Honor of Mario Bunge. 1982 ISBN 90-277-1262-X; Pb 90-277-1263-8
68. W. Krajewski (ed.): *Polish Essays in the Philosophy of the Natural Sciences.* Translated from Polish and edited by R.S. Cohen and C.R. Fawcett. 1982
ISBN 90-277-1286-7; Pb 90-277-1287-5
69. J.H. Fetzer: *Scientific Knowledge.* Causation, Explanation and Corroboration. 1981
ISBN 90-277-1335-9; Pb 90-277-1336-7
70. S. Grossberg: *Studies of Mind and Brain.* Neural Principles of Learning, Perception, Development, Cognition, and Motor Control. 1982 ISBN 90-277-1359-6; Pb 90-277-1360-X
71. R.S. Cohen and M.W. Wartofsky (eds.): *Epistemology, Methodology, and the Social Sciences.* 1983. ISBN 90-277-1454-1
72. K. Berka: *Measurement.* Its Concepts, Theories and Problems. Translated from Czech. 1983
ISBN 90-277-1416-9
73. G.L. Pandit: *The Structure and Growth of Scientific Knowledge.* A Study in the Methodology of Epistemic Appraisal. 1983 ISBN 90-277-1434-7
74. A.A. Zinov'ev: *Logical Physics.* Translated from Russian. Edited by R.S. Cohen. 1983
[*see also* Volume 9] ISBN 90-277-0734-0
75. G-G. Granger: *Formal Thought and the Sciences of Man.* Translated from French. With and Introduction by A. Rosenberg. 1983 ISBN 90-277-1524-6
76. R.S. Cohen and L. Laudan (eds.): *Physics, Philosophy and Psychoanalysis.* Essays in Honor of Adolf Grünbaum. 1983 ISBN 90-277-1533-5
77. G. Böhme, W. van den Daele, R. Hohlfeld, W. Krohn and W. Schäfer: *Finalization in Science.* The Social Orientation of Scientific Progress. Translated from German. Edited by W. Schäfer. 1983 ISBN 90-277-1549-1
78. D. Shapere: *Reason and the Search for Knowledge.* Investigations in the Philosophy of Science. 1984 ISBN 90-277-1551-3; Pb 90-277-1641-2
79. G. Andersson (ed.): *Rationality in Science and Politics.* Translated from German. 1984
ISBN 90-277-1575-0; Pb 90-277-1953-5
80. P.T. Durbin and F. Rapp (eds.): *Philosophy and Technology.* [*Also* Philosophy and Technology Series, Vol. 1] 1983 ISBN 90-277-1576-9
81. M. Marković: *Dialectical Theory of Meaning.* Translated from Serbo-Croat. 1984
ISBN 90-277-1596-3
82. R.S. Cohen and M.W. Wartofsky (eds.): *Physical Sciences and History of Physics.* 1984.
ISBN 90-277-1615-3

Boston Studies in the Philosophy of Science

83. É. Meyerson: *The Relativistic Deduction.* Epistemological Implications of the Theory of Relativity. Translated from French. With a Review by Albert Einstein and an Introduction by Milić Čapek. 1985 ISBN 90-277-1699-4
84. R.S. Cohen and M.W. Wartofsky (eds.): *Methodology, Metaphysics and the History of Science.* In Memory of Benjamin Nelson. 1984 ISBN 90-277-1711-7
85. G. Tamás: *The Logic of Categories.* Translated from Hungarian. Edited by R.S. Cohen. 1986 ISBN 90-277-1742-7
86. S.L. de C. Fernandes: *Foundations of Objective Knowledge.* The Relations of Popper's Theory of Knowledge to That of Kant. 1985 ISBN 90-277-1809-1
87. R.S. Cohen and T. Schnelle (eds.): *Cognition and Fact.* Materials on Ludwik Fleck. 1986 ISBN 90-277-1902-0
88. G. Freudenthal: *Atom and Individual in the Age of Newton.* On the Genesis of the Mechanistic World View. Translated from German. 1986 ISBN 90-277-1905-5
89. A. Donagan, A.N. Perovich Jr and M.V. Wedin (eds.): *Human Nature and Natural Knowledge.* Essays presented to Marjorie Grene on the Occasion of Her 75th Birthday. 1986 ISBN 90-277-1974-8
90. C. Mitcham and A. Hunning (eds.): *Philosophy and Technology II.* Information Technology and Computers in Theory and Practice. [*Also* Philosophy and Technology Series, Vol. 2] 1986 ISBN 90-277-1975-6
91. M. Grene and D. Nails (eds.): *Spinoza and the Sciences.* 1986 ISBN 90-277-1976-4
92. S.P. Turner: *The Search for a Methodology of Social Science.* Durkheim, Weber, and the 19th-Century Problem of Cause, Probability, and Action. 1986. ISBN 90-277-2067-3
93. I.C. Jarvie: *Thinking about Society.* Theory and Practice. 1986 ISBN 90-277-2068-1
94. E. Ullmann-Margalit (ed.): *The Kaleidoscope of Science.* The Israel Colloquium: Studies in History, Philosophy, and Sociology of Science, Vol. 1. 1986 ISBN 90-277-2158-0; Pb 90-277-2159-9
95. E. Ullmann-Margalit (ed.): *The Prism of Science.* The Israel Colloquium: Studies in History, Philosophy, and Sociology of Science, Vol. 2. 1986 ISBN 90-277-2160-2; Pb 90-277-2161-0
96. G. Márkus: *Language and Production.* A Critique of the Paradigms. Translated from French. 1986 ISBN 90-277-2169-6
97. F. Amrine, F.J. Zucker and H. Wheeler (eds.): *Goethe and the Sciences: A Reappraisal.* 1987 ISBN 90-277-2265-X; Pb 90-277-2400-8
98. J.C. Pitt and M. Pera (eds.): *Rational Changes in Science.* Essays on Scientific Reasoning. Translated from Italian. 1987 ISBN 90-277-2417-2
99. O. Costa de Beauregard: *Time, the Physical Magnitude.* 1987 ISBN 90-277-2444-X
100. A. Shimony and D. Nails (eds.): *Naturalistic Epistemology.* A Symposium of Two Decades. 1987 ISBN 90-277-2337-0
101. N. Rotenstreich: *Time and Meaning in History.* 1987 ISBN 90-277-2467-9
102. D.B. Zilberman: *The Birth of Meaning in Hindu Thought.* Edited by R.S. Cohen. 1988 ISBN 90-277-2497-0
103. T.F. Glick (ed.): *The Comparative Reception of Relativity.* 1987 ISBN 90-277-2498-9
104. Z. Harris, M. Gottfried, T. Ryckman, P. Mattick Jr, A. Daladier, T.N. Harris and S. Harris: *The Form of Information in Science.* Analysis of an Immunology Sublanguage. With a Preface by Hilary Putnam. 1989 ISBN 90-277-2516-0
105. F. Burwick (ed.): *Approaches to Organic Form.* Permutations in Science and Culture. 1987 ISBN 90-277-2541-1

Boston Studies in the Philosophy of Science

106. M. Almási: *The Philosophy of Appearances.* Translated from Hungarian. 1989
ISBN 90-277-2150-5
107. S. Hook, W.L. O'Neill and R. O'Toole (eds.): *Philosophy, History and Social Action.* Essays in Honor of Lewis Feuer. With an Autobiographical Essay by L. Feuer. 1988
ISBN 90-277-2644-2
108. I. Hronszky, M. Fehér and B. Dajka: *Scientific Knowledge Socialized.* Selected Proceedings of the 5th Joint International Conference on the History and Philosophy of Science organized by the IUHPS (Veszprém, Hungary, 1984). 1988 ISBN 90-277-2284-6
109. P. Tillers and E.D. Green (eds.): *Probability and Inference in the Law of Evidence.* The Uses and Limits of Bayesianism. 1988 ISBN 90-277-2689-2
110. E. Ullmann-Margalit (ed.): *Science in Reflection.* The Israel Colloquium: Studies in History, Philosophy, and Sociology of Science, Vol. 3. 1988
ISBN 90-277-2712-0; Pb 90-277-2713-9
111. K. Gavroglu, Y. Goudaroulis and P. Nicolacopoulos (eds.): *Imre Lakatos and Theories of Scientific Change.* 1989 ISBN 90-277-2766-X
112. B. Glassner and J.D. Moreno (eds.): *The Qualitative-Quantitative Distinction in the Social Sciences.* 1989 ISBN 90-277-2829-1
113. K. Arens: *Structures of Knowing.* Psychologies of the 19th Century. 1989
ISBN 0-7923-0009-2
114. A. Janik: *Style, Politics and the Future of Philosophy.* 1989 ISBN 0-7923-0056-4
115. F. Amrine (ed.): *Literature and Science as Modes of Expression.* With an Introduction by S. Weininger. 1989 ISBN 0-7923-0133-1
116. J.R. Brown and J. Mittelstrass (eds.): *An Intimate Relation.* Studies in the History and Philosophy of Science. Presented to Robert E. Butts on His 60th Birthday. 1989
ISBN 0-7923-0169-2
117. F. D'Agostino and I.C. Jarvie (eds.): *Freedom and Rationality.* Essays in Honor of John Watkins. 1989 ISBN 0-7923-0264-8
118. D. Zolo: *Reflexive Epistemology.* The Philosophical Legacy of Otto Neurath. 1989
ISBN 0-7923-0320-2
119. M. Kearn, B.S. Philips and R.S. Cohen (eds.): *Georg Simmel and Contemporary Sociology.* 1989 ISBN 0-7923-0407-1
120. T.H. Levere and W.R. Shea (eds.): *Nature, Experiment and the Science.* Essays on Galileo and the Nature of Science. In Honour of Stillman Drake. 1989 ISBN 0-7923-0420-9
121. P. Nicolacopoulos (ed.): *Greek Studies in the Philosophy and History of Science.* 1990
ISBN 0-7923-0717-8
122. R. Cooke and D. Costantini (eds.): *Statistics in Science.* The Foundations of Statistical Methods in Biology, Physics and Economics. 1990 ISBN 0-7923-0797-6
123. P. Duhem: *The Origins of Statics.* Translated from French by G.F. Leneaux, V.N. Vagliente and G.H. Wagner. With an Introduction by S.L. Jaki. 1991 ISBN 0-7923-0898-0
124. H. Kamerlingh Onnes: *Through Measurement to Knowledge.* The Selected Papers, 1853-1926. Edited and with an Introduction by K. Gavroglu and Y. Goudaroulis. 1991
ISBN 0-7923-0825-5
125. M. Čapek: *The New Aspects of Time: Its Continuity and Novelties.* Selected Papers in the Philosophy of Science. 1991 ISBN 0-7923-0911-1
126. S. Unguru (ed.): *Physics, Cosmology and Astronomy, 1300–1700.* Tension and Accommodation. 1991 ISBN 0-7923-1022-5

Boston Studies in the Philosophy of Science

127. Z. Bechler: *Newton's Physics on the Conceptual Structure of the Scientific Revolution.* 1991 ISBN 0-7923-1054-3
128. É. Meyerson: *Explanation in the Sciences.* Translated from French by M-A. Siple and D.A. Siple. 1991 ISBN 0-7923-1129-9
129. A.I. Tauber (ed.): *Organism and the Origins of Self.* 1991 ISBN 0-7923-1185-X
130. F.J. Varela and J-P. Dupuy (eds.): *Understanding Origins.* Contemporary Views on the Origin of Life, Mind and Society. 1992 ISBN 0-7923-1251-1
131. G.L. Pandit: *Methodological Variance.* Essays in Epistemological Ontology and the Methodology of Science. 1991 ISBN 0-7923-1263-5
132. G. Munévar (ed.): *Beyond Reason.* Essays on the Philosophy of Paul Feyerabend. 1991 ISBN 0-7923-1272-4
133. T.E. Uebel (ed.): *Rediscovering the Forgotten Vienna Circle.* Austrian Studies on Otto Neurath and the Vienna Circle. Partly translated from German. 1991 ISBN 0-7923-1276-7
134. W.R. Woodward and R.S. Cohen (eds.): *World Views and Scientific Discipline Formation.* Science Studies in the [former] German Democratic Republic. Partly translated from German by W.R. Woodward. 1991 ISBN 0-7923-1286-4
135. P. Zambelli: *The Speculum Astronomiae and Its Enigma.* Astrology, Theology and Science in Albertus Magnus and His Contemporaries. 1992 ISBN 0-7923-1380-1
136. P. Petitjean, C. Jami and A.M. Moulin (eds.): *Science and Empires.* Historical Studies about Scientific Development and European Expansion. ISBN 0-7923-1518-9
137. W.A. Wallace: *Galileo's Logic of Discovery and Proof.* The Background, Content, and Use of His Appropriated Treatises on Aristotle's *Posterior Analytics.* 1992 ISBN 0-7923-1577-4
138. W.A. Wallace: *Galileo's Logical Treatises.* A Translation, with Notes and Commentary, of His Appropriated Latin Questions on Aristotle's *Posterior Analytics.* 1992 ISBN 0-7923-1578-2
Set (137 + 138) ISBN 0-7923-1579-0
139. M.J. Nye, J.L. Richards and R.H. Stuewer (eds.): *The Invention of Physical Science.* Intersections of Mathematics, Theology and Natural Philosophy since the Seventeenth Century. Essays in Honor of Erwin N. Hiebert. 1992 ISBN 0-7923-1753-X
140. G. Corsi, M.L. dalla Chiara and G.C. Ghirardi (eds.): *Bridging the Gap: Philosophy, Mathematics and Physics.* Lectures on the Foundations of Science. 1992 ISBN 0-7923-1761-0
141. C.-H. Lin and D. Fu (eds.): *Philosophy and Conceptual History of Science in Taiwan.* 1992 ISBN 0-7923-1766-1
142. S. Sarkar (ed.): *The Founders of Evolutionary Genetics.* A Centenary Reappraisal. 1992 ISBN 0-7923-1777-7
143. J. Blackmore (ed.): *Ernst Mach – A Deeper Look.* Documents and New Perspectives. 1992 ISBN 0-7923-1853-6
144. P. Kroes and M. Bakker (eds.): *Technological Development and Science in the Industrial Age.* New Perspectives on the Science–Technology Relationship. 1992 ISBN 0-7923-1898-6
145. S. Amsterdamski: *Between History and Method.* Disputes about the Rationality of Science. 1992 ISBN 0-7923-1941-9
146. E. Ullmann-Margalit (ed.): *The Scientific Enterprise.* The Bar-Hillel Colloquium: Studies in History, Philosophy, and Sociology of Science, Volume 4. 1992 ISBN 0-7923-1992-3
147. L. Embree (ed.): *Metaarchaeology.* Reflections by Archaeologists and Philosophers. 1992 ISBN 0-7923-2023-9
148. S. French and H. Kamminga (eds.): *Correspondence, Invariance and Heuristics.* Essays in Honour of Heinz Post. 1993 ISBN 0-7923-2085-9
149. M. Bunzl: *The Context of Explanation.* 1993 ISBN 0-7923-2153-7

Boston Studies in the Philosophy of Science

150. I.B. Cohen (ed.): *The Natural Sciences and the Social Sciences.* Some Critical and Historical Perspectives. 1994 ISBN 0-7923-2223-1
151. K. Gavroglu, Y. Christianidis and E. Nicolaidis (eds.): *Trends in the Historiography of Science.* 1994 ISBN 0-7923-2255-X
152. S. Poggi and M. Bossi (eds.): *Romanticism in Science.* Science in Europe, 1790–1840. 1994 ISBN 0-7923-2336-X
153. J. Faye and H.J. Folse (eds.): *Niels Bohr and Contemporary Philosophy.* 1994 ISBN 0-7923-2378-5
154. C.C. Gould and R.S. Cohen (eds.): *Artifacts, Representations, and Social Practice.* Essays for Marx W. Wartofsky. 1994 ISBN 0-7923-2481-1
155. R.E. Butts: *Historical Pragmatics.* Philosophical Essays. 1993 ISBN 0-7923-2498-6
156. R. Rashed: *The Development of Arabic Mathematics: Between Arithmetic and Algebra.* Translated from French by A.F.W. Armstrong. 1994 ISBN 0-7923-2565-6
157. I. Szumilewicz-Lachman (ed.): *Zygmunt Zawirski: His Life and Work.* With Selected Writings on Time, Logic and the Methodology of Science. Translations by Feliks Lachman. Ed. by R.S. Cohen, with the assistance of B. Bergo. 1994 ISBN 0-7923-2566-4
158. S.N. Haq: *Names, Natures and Things.* The Alchemist Jābir ibn Ḥayyān and His *Kitāb al-Aḥjār* (Book of Stones). 1994 ISBN 0-7923-2587-7
159. P. Plaass: *Kant's Theory of Natural Science.* Translation, Analytic Introduction and Commentary by Alfred E. and Maria G. Miller. 1994 ISBN 0-7923-2750-0
160. J. Misiek (ed.): *The Problem of Rationality in Science and its Philosophy.* On Popper vs. Polanyi. The Polish Conferences 1988–89. 1995 ISBN 0-7923-2925-2
161. I.C. Jarvie and N. Laor (eds.): *Critical Rationalism, Metaphysics and Science.* Essays for Joseph Agassi, Volume I. 1995 ISBN 0-7923-2960-0
162. I.C. Jarvie and N. Laor (eds.): *Critical Rationalism, the Social Sciences and the Humanities.* Essays for Joseph Agassi, Volume II. 1995 ISBN 0-7923-2961-9
Set (161–162) ISBN 0-7923-2962-7
163. K. Gavroglu, J. Stachel and M.W. Wartofsky (eds.): *Physics, Philosophy, and the Scientific Community.* Essays in the Philosophy and History of the Natural Sciences and Mathematics. In Honor of Robert S. Cohen. 1995 ISBN 0-7923-2988-0
164. K. Gavroglu, J. Stachel and M.W. Wartofsky (eds.): *Science, Politics and Social Practice.* Essays on Marxism and Science, Philosophy of Culture and the Social Sciences. In Honor of Robert S. Cohen. 1995 ISBN 0-7923-2989-9
165. K. Gavroglu, J. Stachel and M.W. Wartofsky (eds.): *Science, Mind and Art.* Essays on Science and the Humanistic Understanding in Art, Epistemology, Religion and Ethics. Essays in Honor of Robert S. Cohen. 1995 ISBN 0-7923-2990-2
Set (163–165) ISBN 0-7923-2991-0
166. K.H. Wolff: *Transformation in the Writing.* A Case of Surrender-and-Catch. 1995 ISBN 0-7923-3178-8
167. A.J. Kox and D.M. Siegel (eds.): *No Truth Except in the Details.* Essays in Honor of Martin J. Klein. 1995 ISBN 0-7923-3195-8
168. J. Blackmore: *Ludwig Boltzmann, His Later Life and Philosophy, 1900–1906.* Book One: A Documentary History. 1995 ISBN 0-7923-3231-8
169. R.S. Cohen, R. Hilpinen and R. Qiu (eds.): *Realism and Anti-Realism in the Philosophy of Science.* Beijing International Conference, 1992. 1996 ISBN 0-7923-3233-4
170. I. Kuçuradi and R.S. Cohen (eds.): *The Concept of Knowledge.* The Ankara Seminar. 1995 ISBN 0-7923-3241-5

Boston Studies in the Philosophy of Science

171. M.A. Grodin (ed.): *Meta Medical Ethics*: The Philosophical Foundations of Bioethics. 1995 ISBN 0-7923-3344-6
172. S. Ramirez and R.S. Cohen (eds.): *Mexican Studies in the History and Philosophy of Science.* 1995 ISBN 0-7923-3462-0
173. C. Dilworth: *The Metaphysics of Science.* An Account of Modern Science in Terms of Principles, Laws and Theories. 1995 ISBN 0-7923-3693-3
174. J. Blackmore: *Ludwig Boltzmann, His Later Life and Philosophy, 1900–1906* Book Two: The Philosopher. 1995 ISBN 0-7923-3464-7
175. P. Damerow: *Abstraction and Representation.* Essays on the Cultural Evolution of Thinking. 1996 ISBN 0-7923-3816-2
176. M.S. Macrakis: *Scarcity's Ways: The Origins of Capital.* A Critical Essay on Thermodynamics, Statistical Mechanics and Economics. 1997 ISBN 0-7923-4760-9
177. M. Marion and R.S. Cohen (eds.): *Québec Studies in the Philosophy of Science.* Part I: Logic, Mathematics, Physics and History of Science. Essays in Honor of Hugues Leblanc. 1995 ISBN 0-7923-3559-7
178. M. Marion and R.S. Cohen (eds.): *Québec Studies in the Philosophy of Science.* Part II: Biology, Psychology, Cognitive Science and Economics. Essays in Honor of Hugues Leblanc. 1996 ISBN 0-7923-3560-0
Set (177–178) ISBN 0-7923-3561-9
179. Fan Dainian and R.S. Cohen (eds.): *Chinese Studies in the History and Philosophy of Science and Technology.* 1996 ISBN 0-7923-3463-9
180. P. Forman and J.M. Sánchez-Ron (eds.): *National Military Establishments and the Advancement of Science and Technology.* Studies in 20th Century History. 1996 ISBN 0-7923-3541-4
181. E.J. Post: *Quantum Reprogramming.* Ensembles and Single Systems: A Two-Tier Approach to Quantum Mechanics. 1995 ISBN 0-7923-3565-1
182. A.I. Tauber (ed.): *The Elusive Synthesis: Aesthetics and Science.* 1996 ISBN 0-7923-3904-5
183. S. Sarkar (ed.): *The Philosophy and History of Molecular Biology: New Perspectives.* 1996 ISBN 0-7923-3947-9
184. J.T. Cushing, A. Fine and S. Goldstein (eds.): *Bohmian Mechanics and Quantum Theory: An Appraisal.* 1996 ISBN 0-7923-4028-0
185. K. Michalski: *Logic and Time.* An Essay on Husserl's Theory of Meaning. 1996 ISBN 0-7923-4082-5
186. G. Munévar (ed.): *Spanish Studies in the Philosophy of Science.* 1996 ISBN 0-7923-4147-3
187. G. Schubring (ed.): *Hermann Günther Graßmann (1809–1877): Visionary Mathematician, Scientist and Neohumanist Scholar.* Papers from a Sesquicentennial Conference. 1996 ISBN 0-7923-4261-5
188. M. Bitbol: *Schrödinger's Philosophy of Quantum Mechanics.* 1996 ISBN 0-7923-4266-6
189. J. Faye, U. Scheffler and M. Urchs (eds.): *Perspectives on Time.* 1997 ISBN 0-7923-4330-1
190. K. Lehrer and J.C. Marek (eds.): *Austrian Philosophy Past and Present.* Essays in Honor of Rudolf Haller. 1996 ISBN 0-7923-4347-6
191. J.L. Lagrange: *Analytical Mechanics.* Translated and edited by Auguste Boissonade and Victor N. Vagliente. Translated from the *Mécanique Analytique, novelle édition* of 1811. 1997 ISBN 0-7923-4349-2
192. D. Ginev and R.S. Cohen (eds.): *Issues and Images in the Philosophy of Science.* Scientific and Philosophical Essays in Honour of Azarya Polikarov. 1997 ISBN 0-7923-4444-8

Boston Studies in the Philosophy of Science

193. R.S. Cohen, M. Horne and J. Stachel (eds.): *Experimental Metaphysics.* Quantum Mechanical Studies for Abner Shimony, Volume One. 1997 ISBN 0-7923-4452-9
194. R.S. Cohen, M. Horne and J. Stachel (eds.): *Potentiality, Entanglement and Passion-at-a-Distance.* Quantum Mechanical Studies for Abner Shimony, Volume Two. 1997 ISBN 0-7923-4453-7; Set 0-7923-4454-5
195. R.S. Cohen and A.I. Tauber (eds.): *Philosophies of Nature: The Human Dimension.* 1997 ISBN 0-7923-4579-7
196. M. Otte and M. Panza (eds.): *Analysis and Synthesis in Mathematics.* History and Philosophy. 1997 ISBN 0-7923-4570-3
197. A. Denkel: *The Natural Background of Meaning.* 1999 ISBN 0-7923-5331-5
198. D. Baird, R.I.G. Hughes and A. Nordmann (eds.): *Heinrich Hertz: Classical Physicist, Modern Philosopher.* 1999 ISBN 0-7923-4653-X
199. A. Franklin: *Can That be Right?* Essays on Experiment, Evidence, and Science. 1999 ISBN 0-7923-5464-8
200. Reserved
201. Reserved
202. Reserved
203. B. Babich and R.S. Cohen (eds.): *Nietzsche, Theories of Knowledge, and Critical Theory.* Nietzsche and the Sciences I. 1999 ISBN 0-7923-5742-6
204. B. Babich and R.S. Cohen (eds.): *Nietzsche, Epistemology, and Philosophy of Science.* Nietzsche and the Science II. 1999 ISBN 0-7923-5743-4
205. R. Hooykaas: *Fact, Faith and Fiction in the Development of Science.* The Gifford Lectures given in the University of St Andrews 1976. 1999 ISBN 0-7923-5774-4
206. M. Fehér, O. Kiss and L. Ropolyi (eds.): *Hermeneutics and Science.* 1999 ISBN 0-7923-5798-1
207. R.M. MacLeod (ed.): *Science and the Pacific War.* Science and Survival in the Pacific, 1939-1945. 1999 ISBN 0-7923-5851-1
208. I. Hanzel: *The Concept of Scientific Law in the Philosophy of Science and Epistemology.* A Study of Theoretical Reason. 1999 ISBN 0-7923-5852-X
209. G. Helm; R.J. Deltete (ed./transl.): *The Historical Development of Energetics.* 1999 ISBN 0-7923-5874-0
210. A. Orenstein and P. Kotatko (eds.): *Knowledge, Language and Logic.* Questions for Quine. 1999 ISBN 0-7923-5986-0
211. R.S. Cohen and H. Levine (eds.): *Maimonides and the Sciences.* 2000 ISBN 0-7923-6053-2
212. H. Gourko, D.I. Williamson and A.I. Tauber (eds.): *The Evolutionary Biology Papers of Elie Metchnikoff.* 2000 ISBN 0-7923-6067-2
213. S. D'Agostino: *A History of the Ideas of Theoretical Physics.* Essays on the Nineteenth and Twentieth Century Physics. 2000 ISBN 0-7923-6094-X

Also of interest:

R.S. Cohen and M.W. Wartofsky (eds.): *A Portrait of Twenty-Five Years Boston Colloquia for the Philosophy of Science, 1960-1985.* 1985 ISBN Pb 90-277-1971-3

Previous volumes are still available.

KLUWER ACADEMIC PUBLISHERS – DORDRECHT / BOSTON / LONDON

MIX
Papier aus verantwortungsvollen Quellen
Paper from responsible sources
FSC® C105338

If you have any concerns about our products,
you can contact us on
ProductSafety@springernature.com

In case Publisher is established outside the EU,
the EU authorized representative is:
Springer Nature Customer Service Center GmbH
Europaplatz 3, 69115 Heidelberg, Germany

Printed by Libri Plureos GmbH
in Hamburg, Germany